Lecture Notes in Computer Science 7216

Antonis M. Hadjiantonis Burkhard Stiller (Eds.)

Telecommunication Economics

Selected Results of the COST Action IS0605 Econ@Tel

Volume Editors

Antonis M. Hadjiantonis
University of Cyprus
Kios Research Center
P.O. Box 20537, 1678 Nicosia, Cyprus
E-mail: antonish@ucy.ac.cy

Burkhard Stiller
Universität Zürich
Institut für Informatik (IFI)
Binzmühlestrasse 14, 8050 Zürich, Switzerland
E-mail: stiller@ifi.uzh.ch

ISSN 0302-9743 e-ISSN 1611-3349
ISBN 978-3-642-30381-4 e-ISBN 978-3-642-30382-1
DOI 10.1007/978-3-642-30382-1
Springer Heidelberg Dordrecht London New York

Library of Congress Control Number: 2012937595

CR Subject Classification (1998): C.2, K.4.1, K.6

LNCS Sublibrary: SL 5 – Computer Communication Networks
and Telecommunications

Acknowledgement and Disclaimer
The work published in this book is supported by the European Union under the EU RTD Framework Programme and especially the COST Action IS0605 Econ@Tel "A Telecommunications Economics COST Network". The book reflects only the author's views. Neither the COST Office nor any person acting on its behalf is responsible for the use, which might be made of the information contained in this publication. The COST Office is not responsible for external Web sites referred to in this publication.

This book may be cited as: COST Action IS0605 Econ@Tel - A Telecommunication Economics COST Network.

Typesetting: Camera-ready by author, data conversion by Scientific Publishing Services, Chennai, India

Printed on acid-free paper

Springer is part of Springer Science+Business Media (www.springer.com)

Preface

The COST Action IS0605 Econ@Tel "A Telecommunications Economics COST Network" was managed within the "Individuals, Societies, Cultures and Health" (ISCH) domain and had a four-year duration between October 2007 and October 2011. COST defines an intergovernmental framework for European Cooperation in Science and Technology, allowing the coordination of nationally funded research on a European level. COST approves specific Actions, which establish networks of such nationally funded research projects, thereby contributing to the reduction of fragmentation in European research investments and opening the European research area to cooperation worldwide.

The goal of Econ@Tel was to develop a strategic research and training network among key people and organizations in order to enhance Europe's competence in the field of telecommunications economics. In that sense, Econ@Tel planned to support related research and development initiatives, and planned to provide guidelines as well as recommendations to European stakeholders for the provision of new converged broadband, wireless, content delivery networks to citizens and enterprises, addressing technology, economics, policies, and regulation.

Therefore, Econ@Tel started a cross-disciplinary work and coordinated in selected fields the development of research methodologies and tools from engineering, media, and business research. By mobilizing a "critical mass" and diversity of European economists, business research experts, engineers, and scientists working in communications, content economics, policy, and regulations it was the first cross-disciplinary COST Action tackling jointly these issues in particular. Thus, Econ@Tel brought together experts from several European member states and associated countries, including Austria, Belgium, Cyprus, Denmark, Finland, France, Georgia, Germany, Greece, Hungary, Israel, Italy, The Netherlands, Poland, Portugal, Romania, Slovenia, Spain, Sweden, Switzerland, and the UK.

The team of Econ@Tel researchers met in the course of the Action in eight Management Committee Meetings (MCM), which covered between one- or two-and-a-half-day technical presentations for all Action members and the public. Additionally, many Working Group (WG) meetings were either collocated with these MCMs or organized at separate times and locations. These WG meetings typically focused on a dedicated topic and covered intense discussions based on presentations given and ideas exchanged. Besides those events, Econ@Tel organized two public workshops in Stockholm and Budapest offering to the wider research public insights into the work performed in the context of Econ@Tel and outcomes achieved. Furthermore, the Action organized two training schools mainly for junior researchers, one in Copenhagen on "Techno-economics of Mobile Networks and the Internet" and the second one in Budapest on "Wireless LTE (Long Term Evolution) Business Models," which distributed Econ@Tel expertise. Finally, 16 Short Term Scientific Missions (STSM) were organized to host an expert, a senior or junior researchers from one institution, at another institution for about one week and to foster the detailed interchange and joint work on common interests. All of those publicly available abstracts, reports, and presentations can be downloaded from Econ@Tel's website at http://www.cost605.org.

Therefore, this volume of the *Lecture Notes in Computer Science* series on "Telecommunications Economics" contains a collaborative and selected outcome of Econ@Tel members' work, who have participated in the COST Action IS0605 "A Telecommunications Economics COST Network." The Action's four WGs are reflected in four major research areas — each of them represented in a main chapter of this book, covering selected results — and especially addressing (a) Evolution and Regulation of Communication Ecosystems, (b) Social and Policy Implications of Communication Technologies, (c) Economics and Governance of Future Networks, and (d) Future Networks Management Architectures and Mechanisms.

Since COST Actions like Econ@Tel are precursors of advanced multidisciplinary research, they anticipate and complement many activities of the EU Framework Programs, and constitute a bridge toward scientific communities of emerging technical, economic, policy-related, regulatory, and organizational stakeholder of COST countries. As such and as mentioned above, these Actions and in particular Econ@Tel also increased the mobility of researchers across Europe and fostered the establishment of scientific excellence in these different key domains among junior and senior researchers.

Furthermore, the editors of this volume would like to extend their thanks to the Springer team for the smooth cooperation in finalizing this scientific report. Finally, special gratitude goes to COST, ISCH domain, and those many COST Action IS0605 Management Committee Members who contributed actively in the course of the Action itself and especially to this book.

February 2012

Antonis M. Hadjiantonis
Burkhard Stiller

COST is supported by the EU
RTD Framework Programme

COST European Cooperation
in Science and Technology

Individuals, Societies,
Cultures and Health (ISCH)

Table of Contents

Telecommunication Economics

Evolution and Regulation of Communication Ecosystems

Social and Policy Implications of Communication Technologies

Economics and Governance of Future Networks

Future Networks Management Architectures and Mechanisms

Summary and Conclusions

Introduction to Telecommunication Economics

Antonis M. Hadjiantonis

KIOS Research Center for Intelligent Systems and Networks, University of Cyprus
antonish@ucy.ac.cy

The telecommunications sector has become a dynamic key area for the economic development of the European Union (EU) and remains in constant evolution. Because of intense competition, telecommunications companies are forced to diversify their offers and thus to propose an increasing number of services. However, economic analysis often ignores important technical aspects of telecommunications and may not be aware of new developments. On the other hand, engineering models often ignore economic factors. Thus, the design and deployment of future networks that incorporate new services are subject to uncertainties such as equipment and capacity prices (due to technological innovation), and demand and supply for services (due to competition). Questions surrounding the deployment of telecommunications infrastructure and next generation networks – together with the massive adoption and increasing demand for new services – bring to the forefront hard questions about the future of Telecommunications and the need to address those questions in a multidisciplinary manner. This book on "Telecommunication Economics" fills that gap, presenting the multidisciplinary outcomes of COST Action "Econ@Tel: A Telecommunications Economics COST Network".

A set of scientific objectives was defined and was addressed in the course of Econ@Tel Action:

a) To support a European engineering leadership gained by new sustainable business models in a fully deregulated and diversified demand framework.
b) To study and identify business opportunities throughout the value chain, especially for enterprises, content, and specialized services.
c) To contribute to a strategy relative to socio-economic needs by increasing the motivation for deployment of cost effective and flexible solutions using networks and content.
d) To provide guidelines and recommendations for utilizing different types of technologies, as well as to model and quantify necessary actions.

To achieve these objectives, leading researchers with various backgrounds, all working on innovative aspects of techno-economic, social, and regulatory issues, focused in an integrated manner on four main areas. The four Econ@Tel areas pertaining to the broader theme of Telecommunications Economics are listed below, while a chapter of this volume is dedicated to each one.

Evolution and Regulation of Communication Ecosystems: This chapter provides insights into the changing landscape of regulatory frameworks, extending across the international roaming domain to the mobile search domain. Particularly, Europe's policy position on the mobile search domain is analyzed, taking in mind the evolution of mobile ecosystems and available policy tools. The angle of privacy protection data

A.M. Hadjiantonis and B. Stiller (Eds.): Telecommunication Economics, LNCS 7216, pp. 1–2, 2012.

and competition policy is also examined, in the context of international roaming of data services and emerging need to revisit relevant regulation activities. The chapter also takes a different perspective on competition policies, highlighting on one hand the conflicting interests of stakeholders which may bias regulations and on the other the employment of community trademarks to track industrial dynamics.

Social and Policy Implications of Communication Technologies: This chapter addresses four different aspects of communications, emphasizing their implications on society and as a consequence to economy. The chapter begins with the examination of affordability indicators for mobile communications, presenting comparative results from seven countries to characterize cost barriers of inclusion. Another comparative study identifies the implications from the deployment of electronic and mobile health (e/m-health) services, using a heart tele-monitoring case. The enhancement of social cohesion is also examined, based on the introduction of services over digital television. Finally, the chapter examines the use of communication technologies for distant cooperation and how leadership is affected in such virtual teams.

Economics and Governance of Future Networks: This chapter covers two broad and interrelated areas, namely the domain of public intervention and governance, and the domain of business model evolution. Regarding the former domain, this chapter describes the public intervention and investments promotion in Next Generation Networks (NGN), also covering the issues pertaining to public-private partnerships (PPP) for infrastructure deployment. In relation to the above, the complementary issues of Internet governance and network neutrality are introduced, making the connection with the latter domain of business model evolution. Aspects of environmentally-friendly or "green" ICT are presented, applied to the cases of "green" mobile tariffs and to the transport sector. Finally, evolving business models for fiber infrastructure deployment and software adoption in Telecommunication complete this chapter.

Future Networks Management Architectures and Mechanisms: This chapter provides a collection of forward-looking technical approaches, addressing aspects which affect Telecommunications Economics. The chapter begins with analyzing the economic aspects of Quality of Experience (QoE), examining links between quality as delivered by the network, as perceived by the user, and as valued by the market. Relevant algorithmic aspects of telecommunications competition for converged networks are also presented. In terms of architectures, Economic Traffic Management and Autonomic Management are examined to reduce the operational expenditures of network and service management. Examples mechanisms of decision support in service contract formation and provisioning of services in rural areas complement the above architectures. The chapter ends with two topics on infrastructure protection, emphasizing on one hand the importance of risk management for ICT and on the other the interdependent nature of ICT and Power networks.

Finally, the last chapter provides major lessons learned during the four years of "Econ@Tel" COST Action and concludes this volume of selected research reports from within the Action.

Introduction to Evolution and Regulation of Communication Ecosystems

Sandro Mendonça

Department of Economics, ISCTE – Lisbon University Institute,
and BRU (ISCTE-IUL), Dinâmia-CET (ISCTE-IUL), UECE (ISEG-UTL)
sfm@iscte.pt

The arena of telecommunications, Internet-empowered interaction, and mobile network technology keeps expanding to encapsulate an ever widening galaxy of customized and utility communication services, that is, specialized applications and the web at large. New challenges are on the horizon as the web makes the transition from the wired Internet to a new model based on ubiquitous access underpinned by a vast array of mobile terminals and the cloud infrastructure. The persistently drastic phenomenon of economic and technical change is transforming economic and social life as well as adding to the pressure on the existing regulatory frameworks.

This chapter addresses a cluster of issues related to mobile technology evolution, telecom industry convergence, and the development of the services market. The confluence of these trends creates uncertainty, fast-paced evolution and a growing potential for economic growth. The European Information Society is faced with new opportunities for policy change and this chapter addresses the emerging new context from a variety of perspectives.

Section 2 on the "Evolution and Regulation of Mobile Ecosystems: European Information Society Policies for the Mobile Search Domain" by S. Ramos, J. L. Gómez-Barroso, and C. Feijóo maps the evolution of the mobile ecosystems by focusing on the particular case of mobile search domain. Analyzing the sector from the angles of the supply side, users' demand and supporting infrastructure this section reaches out for a portfolio of policy tools and approaches.

Section 3 on "International Roaming of Data Services: The Need for Regulation" by M. Falch moves on to acknowledge the importance of the existing national regulations in an environment of deep interconnection and market monitoring (from the angle of competition policy and privacy protection data). Special issues involve how to handle localization data, the spread of M-applications to other sectors, and peer-to-peer mobile services. The features raise questions concerning access, pricing, unbundling, and market-structure that cut across the mandates of different agencies and invite a review of regulation activities as a whole.

"Mobile Regulation for the Future" in Section 4 by Z. Kósa stresses that market issues for international roaming services create a techno-economic realm differing in many ways from markets for other mobile and fixed services. The section examines the supply and demand dynamics in the retail and wholesale segments to illustrate the complications that characterize this dimension of the contemporary telecommunications puzzle. It appraises the recent policy life-cycle and observes to what extent it

A.M. Hadjiantonis and B. Stiller (Eds.): Telecommunication Economics, LNCS 7216, pp. 3–4, 2012.

moved away from more or less objective economic evidence toward regulation based on political negotiations between national stances parties with conflicting interests.

Finally, Section 5 on "Trademarks as a Telecommunications Indicator for Industrial Analysis and Policy" by S. Mendonça and R. Fontana offers insights on how new indicators may guide policy in tracking the dynamics of the telecoms sector. This section uses a new database on Community Trade Marks as a source of empirical understanding on the industrial dynamics and competition trends in the telecom sector in the European market.

Evolution and Regulation of Mobile Ecosystems: European Information Society Policies for the Mobile Search Domain

Sergio Ramos[1], José Luis Gómez-Barroso[2], and Claudio Feijóo[1]

[1] Universidad Politécnica de Madrid—CeDInt, Campus de Montegancedo, Spain
[2] Universidad Nacional de Educación a Distancia, Spain
{sramos,cfeijoo}@cedint.upm.es, jlgomez@cee.uned.es

Abstract. After more than a decade of development work and hopes, the usage of mobile Internet has finally taken off. Now, we are witnessing the first signs of evidence what might become the explosion of mobile content and applications that will be shaping the (mobile) Internet of the future. Similar to the wired Internet, search will become very relevant for the usage of mobile Internet. Within the mobile ecosystem framework, this section will discuss if and how intense public action in the mobile search domain should (could) be. Potential actions refer both to 'conventional' and 'non-conventional' regulatory approaches. Public administrations as procurement bodies may leverage services and thus acting as early deployers of applications is an example of a 'conventional' case, while the use of the wealth of public data with high added value in mobile search scenarios would be one of a 'non-conventional' case. The section will present a list of different policy options and analyze their feasibility. These include policy options aimed both at the demand side (user-oriented) and at the supply side (such as innovation-support policies, regulatory policies, industrial-type policies) of mobile search.

Keywords: mobile search, Information Society policies.

1 Introduction

The mobile base will reach nearly 5 billion subscribers worldwide by 2012. By the end of 2013, broadband mobile connections will account for more than half of all connections and 40% of total subscribers are expected to adopt mobile internet. Its emergence will support an explosion of mobile content and applications. Numerous examples could be mentioned: new entertainment content produced and personalized for the mobile environment, productivity applications for mobile workers, or health and education mobile solutions to increase quality of life, since the mobile device is "the mean to harness collective intelligence at the point of inspiration".

Similarly to the wired internet, many of these new mobile web models will require access to data in an orderly and meaningful manner. Search engines, which are already gateways for more than half of the users connecting to the internet, will

A.M. Hadjiantonis and B. Stiller (Eds.): Telecommunication Economics, LNCS 7216, pp. 5–13, 2012.

therefore become (are already becoming) the mean to reach appropriate content and applications, and to provide additional value to services in mobile platforms.

In addition, mobile search has unique features in providing added value in a number of environments. It exploits the fact that mobiles are very personal devices storing and regularly capturing data about the user, like the user's location, contact lists, preferences, etc. This enables context-aware search services in current and future ambient intelligent environments (e.g. making use of wireless sensors and cognitive techniques). In short, search is likely to become equally or even more critical in the mobile domain than in the wired environment.

As a result, mobile search is becoming an attractive expansion market for all types of existing players (web search engine providers, telecom operators, handset suppliers) and newcomers. It is a clash in many respects: different business cultures (ex-monopolists vs. start-ups), governmental influences (highly regulated vs. non-regulated), business models (subscription-based vs. advertising-based) and the relationship between user and service provider (price-based vs. innovation-based).

2 The Mobile Search Ecosystem

The ecosystem metaphor is useful to describe the relationship of a considerable number of players interacting amongst themselves within a given environment and in which none of players controls the system completely; thus, both collaboration and competition occur at the same time. Today's mobile ecosystem is characterized –in general terms– by an increasingly intense competition at the mobile platform level [3], [9]. With respect to previous periods, the focus of the mobile industry has shifted "from single-firm revenue generation towards multi-firm control and interface issues" [1], [2].

Other authors propose a general model for any mobile content or application [6], [7], which can be adapted also to the mobile search case. The roles of players in the mobile search value network can be broadly divided into three main stages: information processing, delivery and capture/use/interaction.

This three-layer structure is typical of ICT ecosystems [8]; it is developed in Figure 1 presenting the main activities that players can adopt. For each of the three main stages their major contributions are presented in light grey boxes. The dark grey boxes highlight activities which could be considered new to mobile search. Figure 1 includes also the different phases of the mobile search evolution (light to dark blue underlying boxes): the initial on-portal approach (left), the subsequent on-device and additional 'input functionalities' (down), the mobile version of web search (right) and, finally, the context-aware search (up).

The mobile search provision to users requires the contribution of players carrying out most of the activities shown. Obviously some of these players will try, and eventually succeed in, integrating as many activities as possible for a tighter control of the value network. This strategy will result in a “platformization” of the ecosystem, in which each player fights to shift the value towards the platform under its control, ideally including a “gatekeeper” role [4]. Existing web search engines, including the necessary adaptations to the mobile environment, are one of these platforms where the gatekeeping role is mainly related to their favorable and unique position to act as an entry point for end-users to retrieve, subscribe and use content and applications [5].

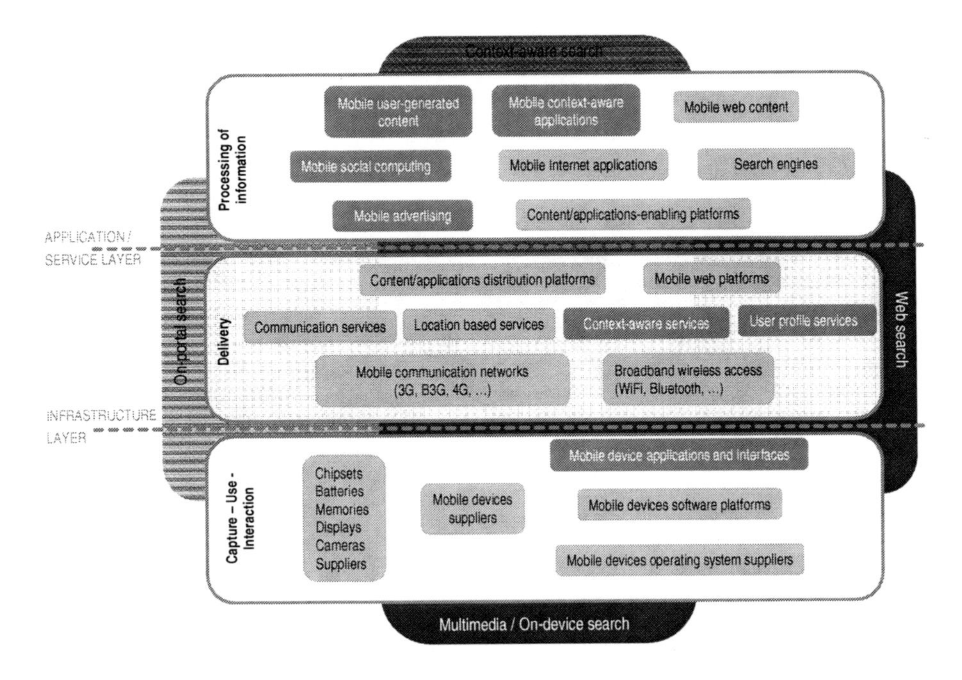

Fig. 1. Activities and players in the mobile search ecosystem [9]

3 Technologies Driving the Evolution of Mobile Search

From a general perspective, there are three main technology families that have a direct impact on mobile search: enabling technologies, search technologies (in general) and specific mobile search technologies, as described in Table 1.

Search technologies, for example for retrieving accurate and enriched content may include semantic approaches, cognitive approaches and multimedia retrieval. Specific mobile search technologies include technologies that render mobile data acquisition, its processing and matching of context-awareness, or that introduce augmented reality technologies to enrich context awareness. Finally, technology components that enable mobile applications include wireless networks (broadband access ubiquity, dynamic spectrum management), sensor networks (Radio Frequency Identification or RFID, Internet of Things), devices (multimedia capabilities, location, interoperability, openness), and cloud computing (web browser, connectivity, security, data protection).

Interestingly, most of these technological building blocks are either already available or in an advanced prototype stage. However, they have not yet been used to any great extent in commercial services and applications. Thus we conclude that, in the short to medium term, there is no missing "critical technological component". Instead, the main technological challenge is to better integrate (existing) technologies. In other words, system integration and technological interoperability is the key to success, rather than the development of new "hard-core" search components.

With regard to the long-term prospects, system integration is particularly challenging, i.e. getting the necessary components operational for the next generation of mobile networks, 4G-type and beyond (arguably the most relevant enabler of mobile search). In addition, current and future networks will also be interoperable with other types of wireless networks such as near field communications for interaction with sensors.

Table 1. Technologies having direct impact on mobile search

	Technology	Keywords
Enabling technologies	Wireless networks	Broadband access ubiquity, dynamic spectrum management
	Sensor networks	RFID, internet of things
	Devices	Multimedia, location, interoperability, openness
	Cloud computing	Web browser, connectivity, security, data protection
Search technologies (general)	Semantic and multimedia	Enriched content search
	Cognitive	Environment understanding
Mobile search technologies (specific)	Context awareness	Context acquisition and processing
	Augmented reality	Enriched context awareness

4 Sources of Revenues in Mobile Search

Using the above framework and compiling revenue models from available literature on the mobile web, applications, content and service models, Table 2 summarizes, from the authors' perspective, the revenue models that mobile search providers are using or could use. They are shown from the perspective of final users, and therefore, intermediate provision models (e.g., white labels, wholesale, brokerage, billing services, software development, hosting, etc) are not considered.

The table includes an example usage scenario (later used for the experts' survey) to better indicate the level of their current existence in practice and their connection with the two main types of mobile search previously described. The fourth column in the table is an indicative of the new business models that context-awareness can bring and which are not present, in general, in the web-based type of search. The revenue models for mobile operators, hardware and software suppliers and other potential intermediaries are not shown in Table 2, although some of them could benefit indirectly from the adoption of mobile search. Finally, also note that the presented revenue schemes are not exclusive and could be complementary to each other.

This list highlights the expectations put on advertising and user profiling as main revenue models in mobile search. In the advertising model, typically the search results are provided free-of-charge to final users and the revenues are generated from third-party advertisers. Advertising models include several very different business strategies. For instance, there could be off-portal campaigns for certain categories of services, such as travel, restaurants, automotive, or consumer electronics. A traditional strategy consists in simply adding a banner on search results, usually including a direct response method as well (a link to a microsite, a click-to-call link, or a short code). This approach fits well, for instance, into events.

As another example, click-to-call text links connected to search results is a simple way to leverage the voice capabilities of mobile devices. Off-portal keyword bidding – also called auctioning-, especially for marketers offering digital content, is another main example. Without exhausting all the possible options, ad campaigns for products related to what mobile operators offer on their mobile portals (ringtones, games, wallpapers, music, video, etc) is an example of on-portal search. Each of these examples could be equally applied to the case of user profiling in exchange for providing the mobile search results.

The list also denotes the still largely unexplored potential of applications where mobile search, typically of the context-aware category, is the engine within. Mobile application providers are looking for business models to incorporate the revenue flow from the application itself, therefore departing from the traditional pay-per-download. There are different business tactics here as well. These can include time-based billing for services, event-based billing for specific situations or item-based billing as a function of the results obtained in the search.

Table 2. Main revenue models for mobile search

Revenue model	Example of scenario of usage	Currently in use in general
Pay-as-you-go (impulse purchase) Subscription	Travel	No
Premium services (basic functionality free)	Leisure	No
Value-added services (additional contract for services on top of conventional ones)	Productivity	No
Merchandising - Affiliation	Well-being	No
Packaged with the mobile device	Consumer good	No
Packaged with the (voice, data) services of the mobile operator	Information	Yes
Packaged with some product or service not related with mobile ICTs	Content	Yes
Advertising in general	Health	No
Advertising linked with product placement	Information	Yes
Exploiting user profile derived from mobile search for marketing purposes	Additional value in purchase of goods	No
Maintained by user community (and free for final users)	Marketing	Yes
Public service (not a commercial one)	Community	Yes
Public service (not a commercial one)	City planning services	No

5 SWOT Analysis of Mobile Search

One of the main objectives of this chapter is to analyze strengths, weaknesses, opportunities and threats (SWOT) for the future success of the mobile search domain (cf. Table 3), as a tool to be considered by European decision-makers.

Table 3. SWOT analysis

EU Strengths	**EU Opportunities**
• Extremely high penetration of mobile technologies and critical mass of advanced mobile users • Industrial landscape strong (operators, suppliers…) and past success stories of co-operation • Main technological puzzle pieces in place (devices, networks, applications…) • Good research standards • Increasingly available and affordable mobile broadband connections • Increasingly available, affordable and usable mobile devices • Availability of content of higher quality for mobile use (geo, land-property registry, …) • Availability of public funded content (broadcasting…) • Multicultural background • Public awareness of privacy issues and increasingly focused laws and regulation	• Improving the integration between web/mobile/PC platform for a richer user experience • Existence of niche markets/services related to mobile search • Mobile search linked with local content (multicultural) • Partner with the experiences of mobile internet usage in developing countries • Be the first to put in place a new (regulatory) framework for API's-interoperability, privacy • Create an open ecosystem for data portability among players and applications based on mobile search • Liberation of European public data for the creation of new services and applications • Use forthcoming disruptions (cloud computing, internet of things…) • Empowerment of the user for granular privacy and identity control internet of things debate.
EU Weaknesses	**EU Threats**
• Techno-economic and market fragmentation (data roaming, standards, application stores, convergent regulation, cultural diversity…) • Need for better / understandable / more secure pricing models and roaming charges in mobile broadband connections • Lack of interoperability and (open) standards • Uncertain strategies for revenue generation, early state of development of business model • Strategic decisions on innovation and investments in (mobile) search are outside the EU • Search mostly dominated by global companies • Lack of entrepreneurship culture and framework for continuing venture capital action	• Delay of enabling technology developments • Increasingly fragmented market (silos, platforms, app stores…) and closed ecosystem (mobile search needs links and references with other domains) • Companies outside EU will control the developments in mobile search • Asymmetry of regulation among electronic communications, internet services and content regimes • Regulatory lag (spectrum management…) • Privacy and data protection issues not acknowledged and solved

On the demand side, Europe enjoys a large base of early adopters of mobile search and a huge mass of mobile users with the economic strength to demand and pay for advanced mobile internet services that satisfy their expectations and requirements. On the supply side, Europe's industry is able to provide users with all the required technology. The industrial tissue is strong and readily available in all required sections of the mobile search ecosystem and particularly strong in some parts of it (telecommunications, handset producers and software and application providers). European companies have significant experience in past success stories (and failures) and, more important, they are increasingly pushed by the market, to simplify mobile tariffs and make them more affordable. Thus, a very positive conclusion is that Europe has both a strong supply and demand side in mobile search. Moreover, European industry is also actively involved in developing countries where mobile devices will become prime means to access the internet. This shared experience could become beneficial in both ways: spreading European innovations and learning from massive usage of mobile internet access.

One specific European asset is that Europe possesses a large collection of high quality information that may trigger advanced mobile search applications at the service of the citizens. Geo-data (e.g., cadastre), images and pictures (e.g., national libraries), or video (e.g., public broadcasters) are examples of data collections in the hands of public authorities, which have already been digitized to a very large extent, that could add significant value to new categories of mobile search. Note that most of this content comes from public sources and/or has been subsidized in the past by public institutions. It seems that public administrations have not yet fully understood how they can exploit in the best possible way this value and how to get into various partnerships and collaborations to unlock its potential. The prospect of "liberation" of public data could also put governments into a favorable position to enforce an open and "loose interoperability" model to allow data portability across applications and players. Forthcoming disruptions in technology could help to deploy such models.

Finally, the many times used but also many times empty-of-practical application motto of "reaping the benefits of Europe's cultural diversity" could become true in the mobile search domain. Some of the most promising applications of mobile search pivot around local information, local culture and specific languages, which is supposed to be complemented by the emergence of many niche markets and services. Civil society is increasingly aware of the need to establish digital identities, which in turn sets the conditions for a stable and firm framework to develop mobile (search) applications, both appealing to users and respectful with them and their preferences and motivations. Europe could be the first to put in place such a light-handed and user-empowered regime shifting the interest of global innovators in mobile search.

Still there are many challenges and barriers to be overcome. The current mobile ecosystem is largely fragmented in terms of both techno-economic models and markets. On the techno-economic side, there are multiple layers (devices hardware and software, applications, networks, development platforms, content platforms, etc) composed of competing, closed and non-interoperable standards. On the market side, the European internal market is far from being established (think on roaming charges, for example) and recent practices (applications stores) keep the tradition of silo incompatible models. Mobile broadband connections are still expensive, particularly in many situations where mobile search would have an extreme value for users (such as finding places in foreign countries), high roaming charges dissuade users from even attempting to connect to the internet. Monetizing mobile search is also still a pending issue. Many business models are possible as discussed before, but none of them has yet crystallized as the winning one.

The mobile search market will remain to be heavily influenced by the web search engines. Given that the most influential ones have all their headquarters abroad, many of the strategic decisions that would influence the evolution of the domain are going to be taken outside Europe's frontiers. To compensate such an effect, a more supportive framework (cultural, institutional and business-like) for entrepreneurs and innovators in Europe would be needed.

The potential delay in the adoption of appropriate regulation regimes (electronic communications, spectrum management, content, consumer protection, etc) will slow the adoption of mobile search. In this sense, a stable, clear and forward-looking framework is desirable which would address the new issues coming from advanced mobile applications.

Finally, there is a risk of a mobile digital divide. Next generation mobile infrastructures may not reach some geographical areas in the short to middle term and the prices both of devices and mobile connections are not affordable for many citizens. Also the skills and physical capabilities to use a mobile device in a search scenario need to be further addressed.

6 Policy Options Scenarios

Following the SWOT analysis conclusions, a list of potential policy options have been considered, grouped in relevant areas of action as shown in Table 4.

Table 4. Summary of potential policy options

Potential Policy Options	Main issues
User-oriented policies aimed at the demand side of mobile search.	• Enhance user-awareness of opportunities and risks. • Create (policy-push) tools for user empowerment.
Innovation-support policies	• Supporting innovators and entrepreneurs through an improvement of the institutional framework. • Promoting living labs, in particular, for mobile applications • Promoting research projects focused on missing technologies
Regulatory policies	• Reforming the mobile search regulatory framework. • Promoting self-regulation of the mobile search industry. • Harmonization and enforcement of EU internal market. • Mandate data portability suitable for mobile search applications. • Creating and enforcing an independent agency.
Industrial-type policies	• Promoting standards and interoperability. • Promoting content production suitable for mobile search. • Supporting a European champion in mobile search. • Setting up a multi-stakeholder platform. • Helping accelerate the deployment of 4G mobile broadband
Public involvement in the supply side of mobile search	• Development of mobile search public services. • Public procurement.

7 Summary and Conclusions

The overall vision on policy is that the mobile search domain requires a combination of different types of actions to thrive and succeed. Looking in detail into each of the potential policy measures, in the first place, there is a need to impel the demand side of mobile search, raising the awareness of users and then empowering them with the tools to manage their data. This should be complemented with reinforcing all polices aimed at innovation: from the support to innovators and entrepreneurs, to the use of living labs and the more traditional research programs. On the regulation side, it is considered that the existing frameworks should be quickly reviewed and adapted to the new needs of advanced mobile applications. However, there is no much faith amongst experts in the self-regulation of the industry or in other actions beyond the regulatory framework like specific agencies or decisions. From the industrial policy perspective, the idea of promoting the use and adoption of open standards and the achievement of a reasonable level of interoperability, including, if needed, a platform to gather all the stakeholders involved is considered of high potential. Helping to develop content for added value mobile search is also highly regarded. Finally, it is thought that for some niche mobile search applications public administrations can have a leading role, setting the conditions for their deployment or even becoming their providers.

References

1. Ballon, P.: Business modelling revisited: the configuration of control and value. Info. 9(5), 6–19 (2007)
2. Ballon, P.: Changing business models for Europe's mobile industry: the impact of alternative wireless technologies. Telematics and Informatics 24(3), 192–205 (2007)
3. Ballon, P.: Control and Value in Mobile Communications: A Political Economy of the Reconfiguration of Business Models in the European Mobile Industry. Vrije Universiteit, Brussel (2009)
4. Ballon, P.: The platformisation of the European mobile industry. In: European Communications Policy Research, EuroCPR (2009)
5. Ballon, P., Walravens, N., Spedalieri, A., Venezia, C.: Towards platform business models for mobile network operators. In: 19th European Regional ITS Conference (2008)
6. Feijóo, C., Maghiros, I., Abadie, F., Gomez-Barroso, J.: Exploring a heterogeneous and fragmented digital ecosystem: mobile content. Telematics & Informatics 26(3), 282–292 (2009)
7. Feijóo, C., Maghiros, I., Bacigalupo, M., Abadie, F., Compañó, R., Pascu, C.: Content and applications in the mobile platform: on the verge of an explosion. Institute for Prospective Technological Studies (2009)
8. Fransman, M.: The new ICT ecosystem. Implications for Europe, Edinburgh, Kokoro (2007)
9. Ramos, S., Feijóo, C., González, A., Rojo, D., Gómez-Barroso, J.: Barriers to widespread use of mobile Internet in Europe. An overview of the new regulatory framework market competition analysis. The Journal of the Communications Networks 3(3), 76–83 (2004)

International Roaming of Mobile Services: The Need for Regulation

Morten Falch

CMI, Aalborg University Copenhagen, Denmark
falch@cmi.aau.dk

Abstract. This section discusses the need for regulation of international roaming charges. This is done through analysis of the EU experiences by a heavy handed price regulation of roaming services.

Keywords: International roaming, EU regulation, regulation of mobile services.

1 Introduction

High international roaming charges have been a matter of concern in several countries. The Australian Competition and Consumer Commission concluded in 2005 that both wholesale and retail roaming charges are too high, but have not turned into direct price regulation. The Arab Regulators' Network is working on different models for introducing price regulation in the Arab region.

Regulation of international roaming is more complicated than regulation of other telecom services for two reasons. First, the market structures on mobile markets are different than on markets for fixed services. Second, regulation of international roaming is difficult to implement at the national level as operators from more than one country are involved.

Markets for fixed services are dominated by incumbent operators having their own fixed infrastructures. Regulatory intervention demanding open access to these networks will benefit new entrants and promote competition at least in the short term. In mobile markets the situation is a bit different as there are more mobile infrastructures on each market. It is therefore less obvious what market implications will be, if a similar kind of obligation is posed on mobile networks.

Partly for these reasons, no regulatory intervention in international roaming charges within the EU area was made before a legislation addressing this specific issue was adopted in June 2007. Although it was recognized that the roaming issue included data as well as voice services, only voice was addressed in the first legislation. Roaming regulation of data was seen to be an even more complicated issue, which needed more analysis before action could be taken.

According to the New Regulatory Framework (NRF), international roaming is defined as a separate market, but the national market studies prepared according the guidelines of NRF did not lead to price regulation in the EU member states. Regulation of this market seems to be more difficult to handle by national regulators than regulation of markets for other telecom services.

A.M. Hadjiantonis and B. Stiller (Eds.): Telecommunication Economics, LNCS 7216, pp. 14–21, 2012.

The legislation on regulation of international roaming made by the EU introduces price caps in both retail and wholesale markets for international roaming of voice services [10]. The major argument for such a heavy handed regulation is that international roaming prices hitherto have been way above cost based prices, and that roaming charges represent a major barrier towards growth in international mobile communication within the EU. The proposal resulted in a reduction in international roaming prices with more than 50%. The question is how this intervention will affect the market in the long term. Will a temporary price cap be enough to keep prices down or will it be necessary to introduce a more permanent type of regulation? Another question is if such regulation can be made without having a negative impact on competition and innovation.

2 Techno-Economic Analysis of Roaming Costs

This section will look at the demand and supply conditions on the roaming market from a techno-economic point of view, and discuss the how the current experiences on regulation of international roaming of voice services can be applied in the design of regulation of data roaming:

- Mobile origination (MO)
- Mobile/Fixed termination (MT/FT)
- International Transit (IT)
- Roaming specific costs (RSC)

Mobile origination includes the cost of initiating the call and for connecting the caller's mobile terminal with the core network. Mobile termination includes the cost related to bring the call from the receiver's core network to his/hers mobile terminal. The international transit costs include the costs related to connect the callers and the receiver's core networks. Roaming specific costs include costs directly related to the roaming transaction. This is the kind of costs, which are not included in a normal international call.

The costs of mobile origination are comparable to those of mobile termination. For data services it may even be difficult to distinguish between origination and termination. For voice mobile termination rates are subject to regulation within the EU and are in principle cost-based. Mobile termination rates per minute varied in October 2007 before regulation was implemented between € 0.0206 in Cyprus and € 0.1882 in Bulgaria. However in most countries the rates were close to the EU average of € 0.0967. The European average for local fixed termination is € 0.0083.

International transit costs depend on the inter-operator tariffs agreed between operators. These tariffs are confidential, but the costs are far below the cost of origination and termination.

3 Structure of the Market for Roaming Services

The market for international for international roaming services differs in many ways from the markets for other mobile services. The special characteristics of this market have been studied in a number of reports and research papers, e.g., [9]. A formal mathematical model of the market is presented in [8] and further extended [7]. The Commission has also created a model in their impact assessment report [1].

3.1 Demand Conditions – Retail

According to the impact assessment report international roaming is used by at least 147 million EU citizens, of whom 110 million are business customers, while 37 million are travelling for leisure purposes [1].

Although international roaming is an important service used by a large group of customers, most subscribers use this service only occasionally, and the level of roaming charges are relatively unimportant for the total cost paid by a mobile subscriber. Therefore only few subscribers will let their choice of operator depend on the roaming charges offered.

Another characteristic of roaming services is that it involves use of many different visiting networks, so even high volume users may only gain a minor benefit by reductions in wholesale roaming charges by a particular operator.

An alternative to international roaming is to acquire a local SIM-card and in this way to avoid paying roaming charges. This solution is in particular attractive for placing local calls abroad, and is used by many potential high volume users of international roaming.

Even though the level of roaming charges is not used as a parameter in competition, it might still be a price elastic service. According to a European fieldwork on roaming prepared in 2006, the main reason for using the mobile phone less while travelling abroad is excessive costs, and six out of ten Europeans would use their phone abroad if prices were more attractive [3].

The impact assessment report assumes in its analysis price elasticity between -0.55 and -1.20 [1]. If this is correct, there will be substantial welfare gains associated with price reductions towards costs based prices (If elasticity are -1.20, even operators revenue will increase).

It should however be noted that the majority of the subscribers using international roaming are business customers, which are likely to be less price sensitive than private customers. Another factor, which may lead to low price sensitivity is lack of transparency. If users are unaware of the actual prices, price reductions will not lead to a higher demand. Use of SMS informing about roaming charges sent to subscriber arriving in country makes the prices more transparent and may therefore increase price elasticity.

International roaming is an international service. International roaming is demanded by customers, when they are abroad. The demand for roaming in a certain area depends therefore not on the number of local customers, but on the number of visitors. Roaming may therefore constitute a major share of the traffic and revenue in tourist areas.

3.2 Supply Conditions – Retail

Suppliers at the retail market include all mobile operators at the national market. As with most other telecom services, the retail market is more competitive than the wholesale market as it includes network operators as well as virtual operators and service providers. Although the suppliers are identical to those offering other mobile services, the market seems to be less competitive. At least the mark-ups demanded by the operators are much higher than mark-ups on other mobile services. The reasons for this is that international roaming is bundled with the subscription for domestic mobile services (unless the customer buys a new SIM-card), are that customers consider roaming rates to be of minor importance compared to rates for other services.

3.3 Demand Conditions – Wholesale

International roaming services are demanded by most mobile operators from their retail customers. Only few operators with an international footprint are able to handle part of their roaming within their own network, while others other – including virtual network operators – will need to buy roaming services from local mobile operators.

Operators will charge their retail customers a price covering the wholesale roaming costs plus a mark-up covering various retail costs such as billing and customer handling. The incentive to reduce wholesale costs depends on how price sensitive the retail customers are. In addition, for technical reasons, it is not always possible for the home mobile network operator (HMNO) to choose the visiting mobile network operator (VMNO) offering the lowest price.

3.4 Supply Conditions – Wholesale

Non-roaming international mobile calls can be handled within the same framework as fixed international calls. Before the liberalization, international fixed calls were priced according to the international accounting rates. The system with international accounting rates dates back to 1865, when the predecessor of the International Telecom Union (ITU) was created. The rate determines how much an operator in one country needs to pay for termination of a call in another country. These rates are negotiated on a bilateral basis and are loosely connected to the costs of maintaining end-to end facilities between the two countries [6].

Payments of international roaming calls are organized in a similar way as other international calls: The call is handled by the operator, where the call is originated. If this involves use of services from other operators, these operators are paid wholesale charges by the originating operator.

In international roaming the call is originated in a visited network and the visited network operator (VMNO) will therefore charge the home network operator (HMNO), holding the subscription of the caller, in order to cover its costs. This payment is settled through the Transferred Accounting Procedure (TAP) by use of Inter Operator Tariffs (IOT) [5].

For receiving international roaming call, the situation is slightly different. Here, the caller will pay for the price for termination of the call in the home network of the roaming subscriber. The HMNO will then transfer the call to the VMNO, who will receive the usual mobile termination rate for terminating the call. In addition to this, the HMNO must pay the international transit charge for a call. Thus the IOTs used for originating roaming calls are not used in this case.

The number of suppliers of roaming services is the same as the suppliers of wholesale mobile services in the respective countries. In most countries all mobile network operators are required to provide roaming services to foreign operators. Thus the number suppliers is 3-4 in most of the EU countries, and it could be expected that the level of competition on roaming services should be at a level similar to that for wholesale provision of other mobile services.

However, IOT have not followed the same decreasing price trends as wholesale charges for other mobile services. Up to 1998 IOTs were based on normal network tariffs for local call with a mark-up at 15% [8]. But since then local tariffs have declined due to price competition. IOTs have not been subject to the same kind of competition, operators within the same country use similar IOTs, and if they relate to normal network tariffs they are set according to the highest rates offered.

There are a number of reasons for this:

- Technical: Limitations in the choice of VMNO;
- Tariffs are agreed on reciprocal basis; and
- Demand conditions and lack of transparency.

As noted above, it is not always possible for the HMNO to choose the VMNO with lowest IOT. This implies that suppliers of roaming services cannot gain market shares by reducing their roaming charges. Market shares of VMNOs are independent of their charges, and therefore, there is no incentive to lower charges as it just will decrease revenues.

IOTs remind in many ways international accounting rates. Usually agreements are reciprocal in the sense that IOTs are independent on the direction of the call. In a market where competition is limited, the HMNO and the VMNO have a common interest in keeping IOT at a high level. The VMNO will get higher revenue for providing the roaming, the HMNO will able to transfer the roaming costs to its retail customers, and will therefore not suffer from this. When a roaming call is made in the opposite direction it is the other operator who benefits.

The international accounting rates contributed to keep international telephone charges at an artificial high level for many years, and prices came down only when it became possible for retail customers to choose an alternative international operator offering lower charges.

The final point to mention is the impact demand conditions and lack of transparency. In elastic demand caused by lack of market transparency as well as other factors implies that suppliers are less eager to reduce costs. For instance there is no incentive to solve the technical complications related to choosing the cheapest VMNO, if cost reductions don't lead to increased traffic demand.

4 EU Regulation of International Roaming Charges

The EU regulation includes regulation of wholesale as well as retail charges. The first proposal from the Commission linked the prices paid for international roaming to prices paid by customers for ordinary mobile calls in their home country [2]. This home pricing principle was replaced by a "European Home Market Approach" in the revised proposal, in which the same maximum price limits are applied in all the EU member states. In the final proposal adopted by the Parliament, the concept of a Eurotariff is used for the maximum price that operators are allowed to charges their customers for international roaming calls within the EU area. The Europe-wide maximum tariffs are defined both for wholesale and retail charges.

Table 1. Eurotariff maximum price while abroad (€-cents per minute) [4]

	Making a call	Receiving a call	Sending an SMS	Receiving an SMS
Summer 2009	43	19	11	Free
Summer 2010	39	15	11	Free
Summer 2011	35	11	11	Free

For data services regulation was introduced 1 July 2009. This regulation limits the price for sending a text message while abroad at €0.11. Receiving an SMS in another EU country will remain free of charge. The cost of surfing the web and downloading movies or video programs with a mobile phone while abroad by introducing a maximum wholesale cap of €1 per megabyte downloaded. This limit will be decreased each year. In addition consumers are protected from "bill shocks" by introduction of a cut-off mechanism once the bill reaches €50, unless they choose another cut-off limit.

5 Discussion

Regulation of international roaming is more complicated than regulation of other telecom services for two reasons. First the market structures on mobile markets are different than on markets for fixed services. Markets for fixed services are dominated by one incumbent operator on each market who has its own fixed infrastructure. Regulatory intervention demanding open access to this network will benefit new entrants and promote competition at least in the short term. In mobile markets, the situation is slightly different as more competing mobile infrastructures are available. It is therefore less obvious what the market implications will be, if a similar kind of obligation is imposed on mobile networks. Second, regulation of international roaming is difficult to implement at national level as operators from more than one country are involved.

For these reasons, a common framework for regulation was not adopted at EU level before 2007. International roaming was defined as a separate market in the market definitions applied in the EU regulatory framework, but the implementation of the new telecom regulation package has not led to any intervention on this market at national level. Market studies for this particular market were among the last to be implemented. In August 2006 market analyses for other telecom services had more or less been completed in most countries, but only Finland had made a decision on international roaming; here the conclusion was that the market was competitive. Since then, no other member states have decided to intervene at the market for international roaming. Regulation of the market for international roaming seems to be more difficult for national regulators to handle than regulation of markets for other telecom services.

The proposal for regulation of international roaming put forward by the EU Commission suggested the introduction of price caps in both retail and wholesale markets for international roaming. The major argument for such heavy-handed regulation was that international roaming prices were much higher than cost-based prices, and that roaming charges represented a major barrier towards growth in international mobile communication within the EU.

An interesting aspect of the proposal from the Commission is the use of a European home-market approach, which implies use of common price caps for all EU member states. This implies that determination of price caps are moved from national to European level. This may therefore be seen as a step towards decreasing the power of national telecom authorities and strengthening regulation at EU level. A common price cap improves transparency for consumers, but it may create a situation where operators in high cost countries may have difficulties in covering their costs in full. It may also create strange pricing schemes, where international roaming becomes cheaper than national roaming.

Both wholesale charges and retail charges have been subject to intensive debate. From the beginning operators were very much against any form of regulation, in particular at the retail level. In spite of amble documentation proving excessive rates without any relationship to costs, it is claimed that there is effective competition on the international roaming market.

Also some Governments have been very reluctant towards regulation. In particular in tourist destinations in Southern Europe, international roaming has proved to be an important source of income.

The EU intervention is a compromise between those asking for cost based roaming charges and the interests of operators – particular those operating in major tourist destinations. Nevertheless seen from the consumers' point of view, it is a considerable improvement compared to the former situation, and it was implemented with an impressive speed (less than one year after the proposal from the Commission was published). It is also a move away from regulation based on more or less objective economic evidence towards regulation based on political negotiations between parties with conflicting interests.

References

1. CEC: Impact Assessment of policy options in relation to a Commission Proposal for a Regulation of the European Parliament and of the Council on Roaming on Public Mobile Networks within the Community, Brussels (2006)
2. CEC: Roaming - Fieldwork September - October 2006, Special Eurobarometer European Commission no. 269, Brussels (2007)
3. CEC: How much do you pay when you use your mobile phone abroad?
4. http://ec.europa.eu/information_society/activities/roaming/tariffs/ (accessed November 20, 2011)
5. Gullstrand, C.: Tapping the Potential of Roaming (2007), http://www.gsmworld.com/using/billing/potential.shtml (retrieved January 15 , 2008)

6. ITU: Direction of Traffic, 1996: Trends in International Ttelephone Tariffs, ITU/Telegeography Inc. (1996)
7. Lupi, P., Manenti, F.: Roaming the woods of regulation: public intervention vs firms cooperation in the wholesale international roaming market, Universita‘ degli Studidi Padova (2006)
8. Salsas, R., Koboldt, C.: Roaming free? Roaming network selection and inter-operator tariffs. Information Economics and Policy (16), 497–517 (2006)
9. Stumpf, U.: Prospects for improving competition in mobile roaming. In: 29th TPRC, Alexandria, Virginia, U.S.A., October 27-29 (2001)
10. Sutherland, E.: The Regulation of International Mobile Roaming. INFO (10), No. 1, 13–24 (2008)

Mobile Regulation for the Future

Zsuzsanna Kósa

Department of Telecommunications and Media-informatics,
Faculty of Electrical Engineering and Informatics,
Budapest University of Technology and Economics
kosa@tmit.bme.hu

Abstract. This section shows technology changes in mobile communications field, like adaptive spectrum usage, 4G mobile services, new M-applications. A theoretic model is shown about the joining process of the value chain, and conflicts are shown on vertical integration in both directions. The staged rises of the mobile communications' level lead two main types of regulator activities. As a consequence in the near future there are some steps by the regulatory agency to be done.

Keywords: regulation, mobile communications, foresight, vision, spectrum.

1 Introduction

There is an inevitable trend seen the wireless mobility. The narrow band GSM telephony has spread all over the world. The new wideband mobile technologies, like LTE (Long Term Evolution) are to step into the market. Several applications, like navigation, locally based e-commerce, twitter or intelligent transport systems are running for the essential limited resource, which is the spectrum band.

This section first shows the most relevant technology changes in mobile communications field, like spectrum allocation, long term evolution technology, and some new applications. Then a model is shown about the extension of the value chain. It is also explained, that the vertical integration could cause tensions in both directions. Next we try to understand the role of regulation in normative and empirical way. The future-oriented regulatory tasks are in the next subsection, based on the previous statements. As a consequence of the spreading process, we can see the need for co-regulation with other agencies of related fields.

2 Spectrum Allocation

Spectrum allocations are more and more important. One new band usable for communications purposes has high value for the service providers. To build up a new wireless technology always new frequency bands are needed. The allocation process goes in several national, regional CEPT-ECC (European Conference of Postal and

A.M. Hadjiantonis and B. Stiller (Eds.): Telecommunication Economics, LNCS 7216, pp. 22–32, 2012.

Telecommunications Administrations, Electronic Communications Committee[1]), and global ITU-R (International Telecommunications Union, Radio-communications' sector[2]) level. For service providers the right to use a slice of the spectrum means a high value asset. The national governments spend a lot on frequency allocation processes, and as a consequence, governments have the right to sell the free frequency bands, mainly through auctions, sometimes through beauty contest.

There are new flexible band techniques called Software Defined Radio and Cognitive Radio. According to REPORT ITU-R SM.2152 [4]:

- "Software-defined radio (SDR): A radio transmitter and/or receiver employing a technology that allows the RF operating parameters including, but not limited to, frequency range, modulation type, or output power to be set or altered by software, excluding changes to operating parameters which occur during the normal pre-installed and predetermined operation of a radio according to a system specification or standard."
- "Cognitive radio system (CRS): A radio system employing technology that allows the system to obtain knowledge of its operational and geographical environment, established policies and its internal state; to dynamically and autonomously adjust its operational parameters and protocols according to its obtained knowledge in order to achieve predefined objectives; and to learn from the results obtained."

These techniques do not need preliminary given frequency band. These adaptive technologies provide much more effective shared spectrum usage. So, these technologies are good to serve in huge metropolitan areas, where the density of the users is very high.

3 Long Term Evolution (LTE) Mobile

The next generation mobile technology deployed at the moment is called Long Term Evolution LTE (4G). The official name of these new technology standards, given by ITU is International Mobile Telecommunications Advanced technology [3], [5].

The major feature of LTE mobile internet system is that the upload speed can be much more than in the previous mobile internet systems. The LTE is mainly for data transmission not for voice. The other feature of LTE is that it tries to integrate the existing mobile systems. This may cause faster spread and lower prices at the introduction period.

A new capable technology stepping into the market would produce changes in the whole industrial value chain. The value chain of the broadband mobile services consists of devices, infrastructure services, applications and some solutions. The first application was localization based content providing. The more spread application is the navigation on smart mobile phones. These applications are mainly from e-commerce, and social applications are also spreading.

---

[1] The European organization CEPT-ECC see `http://www.cept.org/ecc/`

[2] The Global organization ITU-R see `http://www.itu.int/`

4 New Applications Based on Mobile Services

The users' prospective has been changed: "digital way of living" - is the new approach, integrating the previously segmented disciplines of e-commerce, e-learning or e-health, e-working. Digital living needs new approach to free and working hours, a new concept of working place, and also the opportunity costs in time and working environment. There are several new applications based on mobile telecommunications services. Most of them are provided by large telecommunications companies or their subsidiary companies.

The first group of applications is *mobile payment*, which is one of micropayments type. The deposit is paid or debt is asked before the transaction, and the payment service goes within the system. Acceptance of mobile payment is similar to other debit cards. In general micropayment systems do not pay interests after the deposit, therefore do not ask fee for the transactions. In case of mobile payment, it mainly goes from credit, so the transaction must have a fee over the exact payment.

The second group of applications and solutions is *mobile health*, including special devices based on mobiles. Some sensors may measure the health conditions of the users, evaluate the data and inform the user and/or other helping assistance. In case of high health risk or in case of emergency, the emergency call could be started without any human interaction. There are other, information services serving health issues: localization-based pharmacies, electronic prescription and data-storage of previous records of laboratory.

The third group of mobile based applications is *m-learning*. The mobile becomes part of the information infrastructure to learning management systems (LMS). This ecosystem consists of few on-line teachers, life-long learners, and schools. Really good improvement of competences may come from the blended learning process including personal consultations with the teacher and the facility based on-line communication through screens. There are three screens to learn: TV screen, computer screen and the mobile screen and online e-learning systems could use at least two of these screens.

E-government sometimes is served *on mobile*. This is the communications channel which reaches almost the whole population. Mobile communications are also available for needy population: jobless, homeless, immigrants, handicapped peoples, old age pensioners. So, the local public services - like help to the labor market, social food distribution, help-services for large families in education - may be organized through mobiles.

Navigation is available for mobile customers through satellite systems, some of mobile producers provide it without further fee. Locality based services can be organized based on mobile devices too. Navigation services are the basis for the future intelligent transport systems of avoiding traffic jams, mobile payment for highways or parking.

We can see that forward integration of mobile service providers is a strong ongoing process. Mobile industry spreads into other sectors, together with new applications serves as a business infrastructure for the business.

5 Joining Process of Value Chains

On the next figure is the joining process of the value chain in case of a new capable technology steps into the market. Four phases can be seen:

- 1st phase - Based on the new technologies the quality rises, the prices diminish, the infrastructure spread into new business areas;
- 2nd phase - Some complementary industries build in their services the already existing infrastructure;
- 3rd phase - The value chains merge, and form a new one together, providing a new compound service for the customers;
- 4th phase- The new *compound service itself becomes* the basis for other industries as a *business infrastructure*.

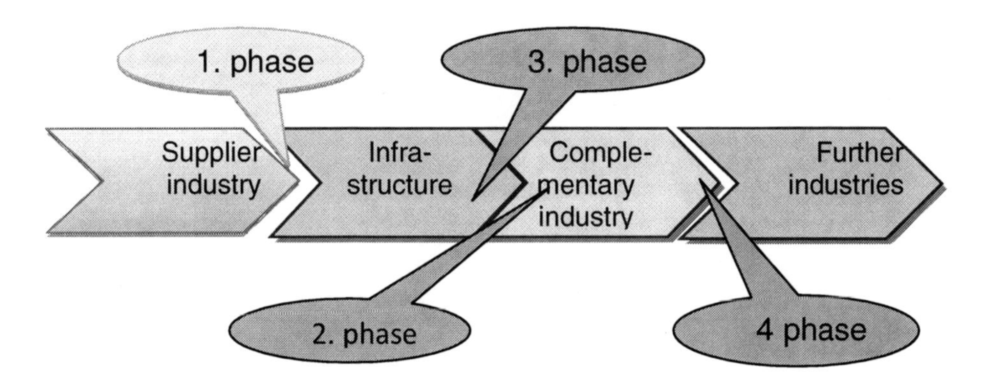

Fig. 1. Value chain extension in case of innovation

As we have seen previously, mobile telecommunications services spread into other sectors. Let us to build up models of this extended service provision. When two sectors are to joining along the value chain, there are always ambitions from both sides to integrate backward and forward the close value producing steps. As you may see on Figure 2, the telecommunications service providers try to integrate new ICT applications to provide more enhanced services for related industries; the professional users of related industries try to extend its own activities backward, using new ICT technologies. Both vertical integration processes of different industries go through tensions, even if it is forward or backward integration. The first reaction of the neighboring industries may be to resist, the second reaction might be to build up strategic alliances, and the third step can be the value chain joining process.

Regulation and the regulatory body in this joining process should play as a defender of the community interest. Community interest can be:

- Let the service provider provide more enhanced services, in order to provide basic services at lower price for the large scale of population;

- Let the innovative new technologies step into the market, in order to have the ICT market combatable and avoid any kind of monopolistic situation.

The regulators' balancing role may be examined also in Figure 2.

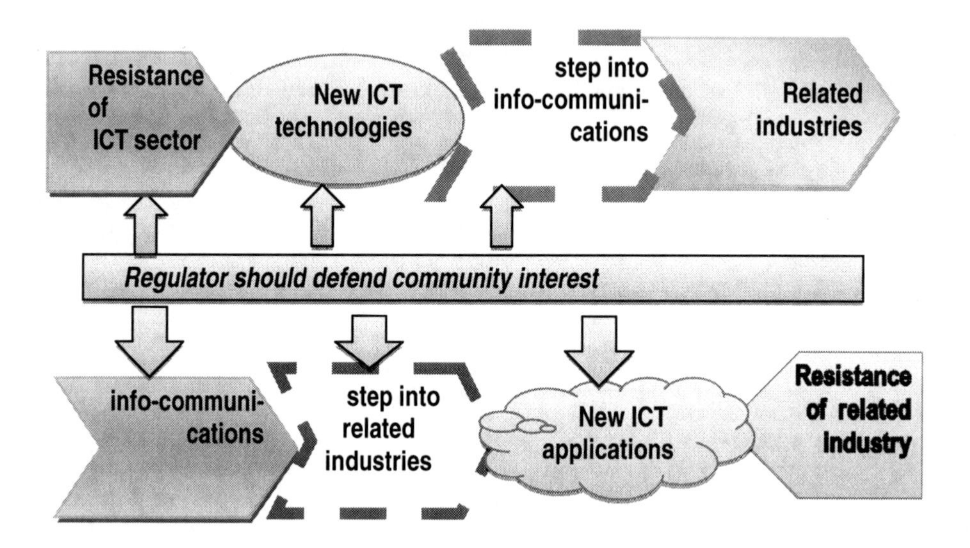

Fig. 2. Backward and forward vertical integration with related industries

6 Role of Regulation

Sector-specific regulation in wider sense includes market monitoring, rule making, individual decisions and information providing about the certain market. It includes also participation in the standardization process, although it is considered a self-regulation of industrial players. It should match to overall market regulation topics, like customer defense, privacy defense or accounting rules.

There are several function expected from a regulator: handle customer dependency, enhance innovation and network development, and handle forward and backward integration processes. Some solidarity steps should be done to ensure universal access to the infrastructure. Let us see the innovation model of the ICMT sector model reflecting [1] ideas on information technology.

We can see *a staged development* of the info-communications services at the "state of the art" level. New technologies come and provide business advantages for the first user groups. Then these technologies spread into the everyday life, and the business advantage melts away. Later this technology becomes essential even to participate in business. Remember the example of mobile telephony: first it used to be a status symbol for high level managers, then became normal usage for workers, now it is an expectation to have even for jobless or homeless peoples. As we see, the participation level shows also a staged development, only with a bit delay.

What should do with this staged development process the regulator? When the scale is rising, should react, in order to enhance the development of the market. There are two main types of regulatory action: innovative and solidarity actions of regulator. In case the state of the art level increases, an innovative action should be done, in case the participative level increases, a solidarity action should be done. See the whole staged development process together on Figure 3.

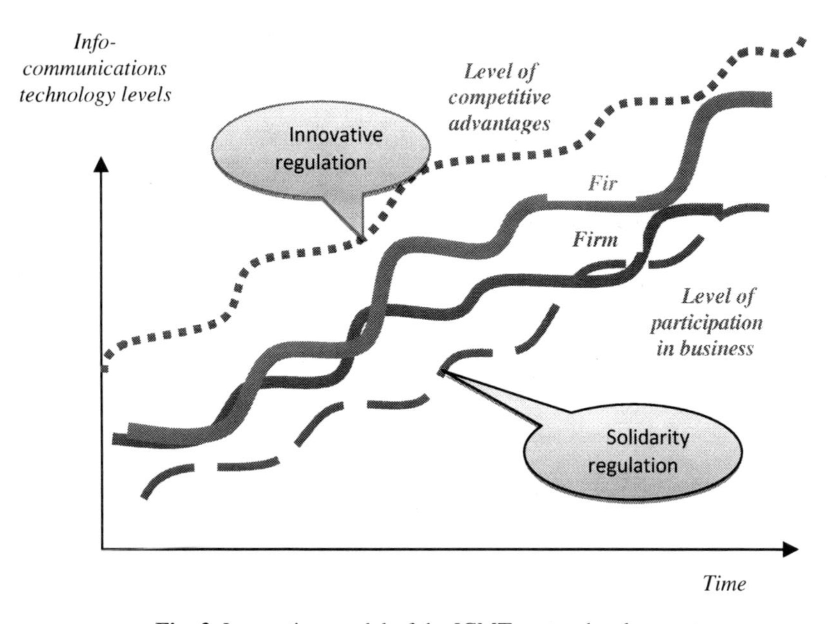

Fig. 3. Innovation model of the ICMT sector development

Innovative regulatory actions are:

- Let a new technology to enter into the market;
- Ensure scared resources (spectrum or identifier) for the new technology;
- Ensure service based competition for innovative applications;
- Participate in standardization process.

Solidarity regulatory actions are:

- Make a price regulation;
- Universal access and universal service regulation;
- Enhance facility based competitions or ladder of investment;
- Ensure local physical networks in a non profit basis.

After having seen the normative approach of the regulation, we can also deal with the regulation with limited rationality [2]. The real regulator is a government agency, with limited power having external stakeholders. In case the value chain extends, the players also change in the stakeholder model of the regulatory agency. So, there will be new players to be fulfilled with regulatory interventions on the market. Figure 4 shows the stakeholder model of the regulatory body in the future.

Based on these – normative and empirical – theoretic models of regulation duties some more detailed real regulatory steps can be described on mobile field in the next subsection.

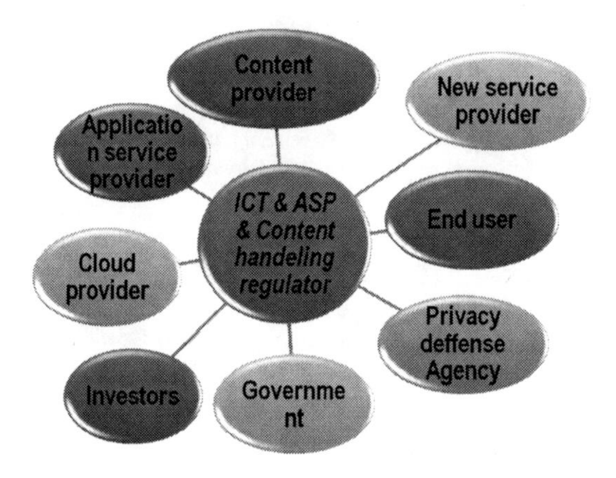

Fig. 4. Stakeholder model of the regulator in the future

7 Future-Oriented Regulation Tasks in Mobile Communications

There are *classical regulation issues* in mobile, as in any kind of matured industries-Termination rates at interconnecting; prices of mobile roaming; domestic roaming in case of a newcomer on the national market; interconnection with internet service providers; prices of data transmission, and universal access provision through mobiles are also to be regulated.

Usually the *value of a spectrum* is based on the economic value of the market presence on the mobile market. It is usually sold by the government on auctions. But there are new technologies without allocated spectrum.

There is a new emerging issue on *frequency management* field. There are free bands without preliminary allocation, called "white spaces", where the usage is shared cognitive radio and software defined radio, as it is mentioned above. This new element of the radio-communications' market would produce the *following two problems* for the national regulatory agencies on the European mobile service market.

- The frequency fee nowadays goes mainly to the state budget of the national governments, and it covers the operational costs of the frequency management institutions, sometimes there is cross-financing to the whole regulatory agency from frequency fees. It would be missing from the state budget and the regulatory agencies would be financed from general taxes from other industries;

- The frequency allocation is one of the tools in government hand, to enforce some content providing issues in case of broadcasting channels. If there is no such tool in hand, it would be very difficult to enforce the expected percentage of the European or national content in the provided entertainment channel programs. It could open the door to an overflow of films or games or some non-wanted contents from out of Europe.

Although the spread of the white spaces is not an interest of European regulators, the trend is coming and it would be very difficult to stop it. It will be one of the discussed topics on the next World Radio Congress of ITU, and it is hard to foresee the results.

WHO issued a precaution approach to *power emission*, and there is a European standard on the maximum specific absorption rate (SAR). But customers do not have the information about the possible danger of the over-usage of mobile their handsets. The regulatory bodies should enforce service providers to inform their customers about radiation effects of mobiles. There are some survey, that the risk is higher in case of young children and persons already having decease.

Enhancing innovation, the *technology neutrality* has been a huge discussion since 2003. We can read on ITU side: It "means that different technologies offering essentially similar services should be regulated in similar manners. However, technologies offering similar services do not necessarily have similar features in all aspects, and exactly identical regulations may, therefore, result in the advantage of one technology over another in the market." [7].

The *reliability of mobile communications* services is essential in case further industries build themselves on it. Other industries are regulated mainly by their own agencies, like financial sector, transport, health and social welfare, privacy defending.

8 Cross-Sectored Co-regulation Issues for Mobiles

The trust building toward mobile communications' reliability can be handled through cross-sectored regulation processes together with different government authorities both at national and at European level. This is slightly different approach to the existing self- and co-regulation approaches, which was mainly industry based or civil based self-regulation [6].

The value chain of the ICMT (Informatics, Communications, Media Technologies) sector is expanding, and form a union together with value chains of other industries. Regulatory activities should follow the process of spreading information technologies. The new approach is to have common responsibility – among different sectors and their regulatory bodies – for the new, integrated network infrastructure for the information based society. Figure 5 shows that regulation should follow these changes in an industrial value chain, and regulation should extend these activities into new fields.

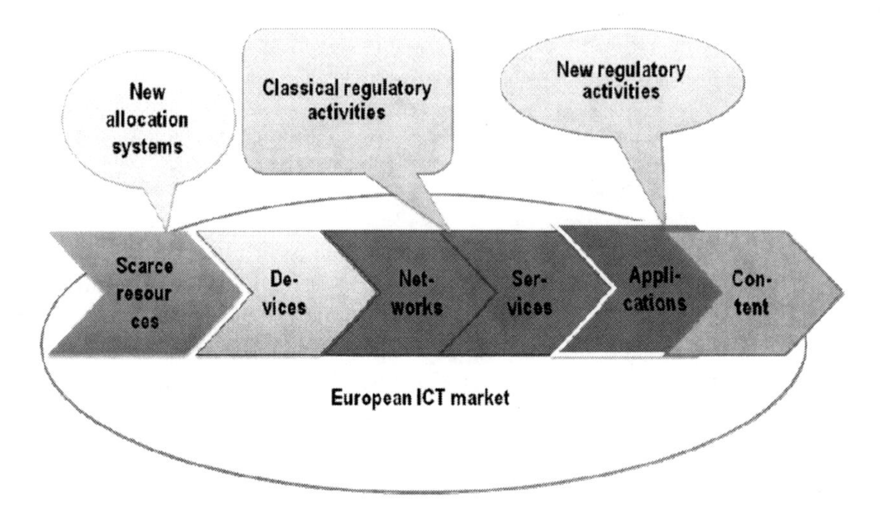

Fig. 5. Extending regulatory activities

These new fields have also been regulated until now by other sector specific agencies, but based on other technologies. There is a need to join the regulatory activities with the agencies of related industries, like financial regulator, transport regulator, privacy defending agency, health care regulator. See the following examples of the joint regulatory activities.

What is the border of payment services for mobiles? These systems (as we have described above) work as any other micropayment systems. Mobile based micropayment goes from debt, like debenture cards. So, the regulation of these services should go together with other financial services. But the sector-specific regulatory body should enforce the accounting separation of the revenue on mobile communications services and micropayment services

There is a regulation of emergency calls (e-calls), which should be similar in every EU (and associated) countries. The introduction of e-calls helps the tourism and the trans-European transport systems, because the caller party should not has to know any local emergency call, or should not has to speak local languages to ask for help in emergency case. The new regulation helps the automated mechanisms to spread. "In case of a crash, an e-call equipped car automatically calls the nearest emergency centre." If we think on the traditional human based emergency calls, local emergency services might be informed a bit later, if there is a previous call centre before them in any cases. So, it is useful let to use both emergency calls: the trans-European and the local ones too. (In case of Hungary besides the European 112 the local 104 should remain too). We can see that the introduction of this new e-call public service is going slower than it was expected before, although there are many possibilities in it in long term. The new cars should be equipped with this ability, but the old ones do not have them. Regulators should force to build the emergency call devices into older vehicles too. This would enhance the trade of mobile industry in a new field.

There is a privacy issue how to *share information on localization data*. In case of e-calls it is already done the exemption in order to help to the person. Other cases should be regulated among other privacy issues, like: localization data of small children for their parents, localization data of employees during working hours to their employers, localization data of assistant peoples at emergency services any time, etc. These are non-criminal cases but there are personal pending situations. On the other hand, service providers should provide localization data sharing services in case the customers want it: old age people living alone may share their localization data with their family members or health assistant services.

There are several *mobile based applications on e-health.* Data sharing of these equipments should be regulated together with privacy defending bodies. How to extend social security systems to m-health is one of the next questions to co-regulate with healthcare authorities. The expected benefit is that the patients may reach the health system earlier, and the rehabilitation could be cheaper. The most important would be to introduce M-health devices for professional drivers of trucks and public transport vehicles, in order to prevent mortal accidents.

The most recent development is a proposal about establishing the *Connecting Europe Facility.* In this, there are common rules of the Transport, Energy and Telecommunications industries. The initiative includes broadband mobile services too. On the other hand, there are expectations in the proposal to build up European high speed backbone networks, so the mobiles could also use that in the future [8].

9 Conclusions

Having seen the relevant regulatory steps in mobile communications' field we can conclude there are several new applications. Based on that, the mobile communication service providers try to make a forward integration, and build up mobile based application services, spreading into new related industries. Regulatory agency should follow the extension of the value chain.

As a result of the spreading ICT sector, the regulatory agency should build up cross-sectored cooperation among related regulatory agencies. Examining the role of regulation we can see, that the normative approach shows two groups of regulative actions: steps for innovations and steps for universal services. The empirical approach to regulatory body shows, that there will be new stakeholders on the screen.

Having the future oriented analysis of the regulation, the regulatory agency may be prepared to the most probable future, avoid the least acceptable scenarios and achieve the most desirable ones.

References

1. Carr, N.G.: IT doesn't Matter. Harvard Business Review, 5–12 (2003), `http://www.scribd.com/doc/14468461/Carr-03-IT-Doesnt-Matter`
2. Fehr, E., Tyran, J.R.: Limited rationality and strategic interaction, University of Zurich working papers no 130 (2002), `http://www.iew.unizh.ch/wp/iewwp130.pdf`

3. ITU: Resolution 56* Naming for International Mobile Telecommunications (2007), http://www.itu.int/pub/R-RES-R.56-2007
4. ITU: Definitions of Software Defined Radio (SDR) and Cognitive Radio System (CRS), REPORT ITU-R SM.2152 (2009), http://www.itu.int/pub/R-REP-SM. 2152-2009
5. ITU: Global Standard for International Mobile Telecommunications 'IMT-Advanced' (2011), http://www.itu.int/ITU-R/index.asp?category=information&rlink=imt-advanced&lang=en
6. Gercke, M., Tropina, T., Sund, C., Lozanova, Y.: The role of ICT regulation in addressing offenses in cyberspace. In: Global Symposium for Regulators, International Telecommunications Union, pp. 28–39 (2010)
7. Koops, B.-J.: Should ICT Regulation be Technology-Neutral? Starting Points for ICT Regulation Deconstructing prevalent Policy One-Liners, IT & law Series. In: Koops, B.-J., Lips, M., Prins, C., Schellekens, M. (eds.), vol. 9, pp. 77–108. T.M.C. Asser Press, The Hague (2006)
8. Koops, B.-J., Bekkers, R.: Interaceptability of telecommunications: Is US and Dutch law prepared for the future? Telecommunications Policy 31, 45–67 (2007)

Trademarks as a Telecommunications Indicator for Industrial Analysis and Policy

Sandro Mendonça

Instituto Universitário de Lisboa (ISCTE-IUL), and BRU (PEst-OE/EGE/UI0315/2011), Dinâmia-CET (ISCTE-IUL), UECE (ISEG-UTL), Portugal
sfm@iscte.pt

Abstract. Innovation and economic transformation are difficult phenomena to measure, especially in high-tech fast changing sectors. Dynamic competition is a key driving force behind growth and a complex, multidimensional nature calls for an integrated approach. Recently a research agenda that see trademark as a valuable addition to the economic indicators arsenal has been gaining momentum. This section makes the case for employing trademarks as an indicator for assessing dynamic competition and international competitiveness in the telecommunications equipment and services sector. It considers community trademark data to explore stylized facts and recent trends in the European market, a sophisticated and contested world sales pitch on which this data can be employed to generate substantive academic insights as well as useful knowledge for policy-makers at the EU and national levels.

Keywords: telecommunications, indicators, trademarks, competitiveness, policy Lisboa, Portugal.

1 Introduction

Today's economy is increasingly a communication-intensive world trade system. Competition isn't just about new technology and better service. It is also a fierce struggle for the consumer's attention in an increasingly overwhelmed information environment. Trademark data can constitute a source for fresh insights on how companies and countries are being able to put their message across in the contemporary innovation-driven global economy.

As Schumpeter suggested innovation is about introducing new or improved goods and services into the market, launching new or improved manufacturing and service delivery systems into economic activity, discovering and fulfilling all together new or modified needs and desires [1]. Innovation is then about converting relevant technical knowledge into working useful products. Yet, while an increasing quantity of research has been invested into the study of the sources and effects of innovative operations, it continues to operate on the basis of narrow empirical resources. Dynamic competition is a complex, multidimensional phenomenon. At the same time innovation and industrial transformation are difficult to measure. Engaging with real phenomenon of competition, as opposed to standard textbook accounts of neoclassical competition, is

A.M. Hadjiantonis and B. Stiller (Eds.): Telecommunication Economics, LNCS 7216, pp. 33–41, 2012.

an important task since it is a key driving force behind economic growth. This brief note attempts to highlight the potential of this new indicator in the context of applied telecommunications economic analysis.

2 Mapping and Measuring the Commercial Creativity of Firms and Nations

Two intellectual property rights share a great deal: patents and trademarks. They yield precious macroeconomic data. Both are "back office" measures of how a country is succeeding in the structural transition to a knowledge-based economy. Both are correlated with innovation performance and provide an insight into ongoing processes of industrial change. Patents are assigned to original, non-trivial and productive inventions. Trademarks such as brand names and logos are solicited by firms to distinguish and protect the reputation of goods, services and their corporate identities. These indicators are also complementary: patent counts are a pointer of technological expertise and trademark statistics are an indicator of commercial capability.

Patents and trademarks are here understood as economic indicators of innovation and industrial change. Numbers of applications of these intellectual property rights allow for complementary readings of the state and evolution of an economy. One key difference sets them apart: trademark data is a much less exploited information resource. Patents have been used for a number of decades now and are commonplace in standard economic benchmarking publications, such as those from the OECD, whereas trademarks have been much less used in applied analysis. They offer the possibility of yielding fresh insights in the process of transformation and competition among modern.

The business of branding goods and services has been an ordinary part of economic life since time immemorial. Symbols were used to keep track the origin of bricks used in Roman constructions, familiar names where used by medieval artisans to distinguish the products of their craft, calves roaming on the western prairies were marked by ranchers to identify their property. However, as a source of empirical insights trademarks have been neglected; in spite that branded goods have become a distinguishing feature of economic life in the twentieth century [2]. This neglect has co-existed, nonetheless, with an increasing awareness in the social sciences that branding (or trademarking) has been emerging as a particular phenomenon worth of explanatory work (e.g., in sociology, psychology), but also of prescriptive consideration (e.g., management studies).

Only very recently, applied economic literature has shown that branding activity may be also captured by trademark applications for the analytical purpose of assessing and monitoring ordinary and extraordinary economic life: trademarks have been shown to provide consistent indications innovation behavior and industrial change. Trademark statistics are interesting because they are *i*) increasingly available on electronic platforms, *ii*) regular long-term data availability, *iii*) broken down by product classes, *iv*) able to capture service and SME innovation, and *v*) close to commercialization of new products.

Trademarks are exclusive rights to distinctive words, symbols, shapes, or even smells and holograms. The legal framework of this IPR has been evolving.

A milestone was the Paris International Convention for the Protection of Industrial Property of 1883 (known as the ''Paris Convention'') covering patents, trademarks, and designs [2].

3 Community Trademarks for Competitiveness Analysis

Marks and other distinctive signs (such as logos, slogans, three-dimensional marks objects) they are commercial signatures that distinguish the products and their producers on the marketplace. The flow of trademarks applications reveals the initiative of creating presence and recognition in targeted commercial arenas. With necessary methodology care data on this activity may be used to build a better understanding concerning rhythms and directions of development of national economies, for instance, in the European context.

One place to look is indeed Europe. With almost 500 million consumers and arguably one the most attractive, and sophisticate markets of the world, this is an especially interesting arena to analyze. Thanks to the Community Trade Mark (CTM), an IPR tool that covers the EU territory and which came into existence in 1996, we can observe trends as they develop. In this section we use for the EU-15 countries' applications obtained from the Office for Harmonization in the Internal Market (OHIM), which is responsible institution for managing CTMs. This European right can be seen as a success story seems it appears very clearly to have provided an answer to a latent demand for this kind of European-wide intellectual property right. The last year in record, 2007, is a special one because the volume of applications became more twice of the level of the launch year, 1996, for the first time [2].

Figure 1 shows absolute numbers for total applications from all EU and non-EU countries since the creation of the launch of the CTM. The first year should be regarded as an abnormal year since intentions to trademark were being accumulated for at least two years before. The local peak noticed around the year 2000 can be associated to the "New Economy" bubble, since trademarks covering high-tech categories where driving growth; only by 2005 aggregate numbers surpass those of the year 2000. By the end of the decade, though, the effects of the so-called "sub-prime crisis"/"great recession"/"little depression" are visible.

4 Trends in Communication-Intensive Knowledge Economy

A total of 250,000 trademarks were submitted for the whole territory of the EU during last year. Applications by EU-based firms and organizations constitute around two thirds of this figure. Within Europe Germany has been by far the country originating the largest number of trademarks, a share of a quarter of total applications among the 27 EU Member States. Figure 2, which depicts absolute numbers of new CTMs requested by the most important countries in terms of volume of applications, shows a clear trend for the last decade. There is a marked and generalized movement towards intensified trademark deployment.

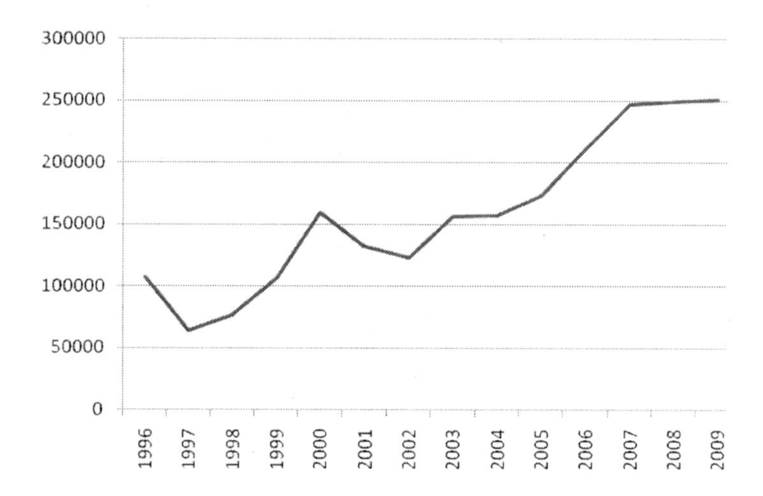

Fig. 1. Total CTM applications, 1996-2009 [3][1]

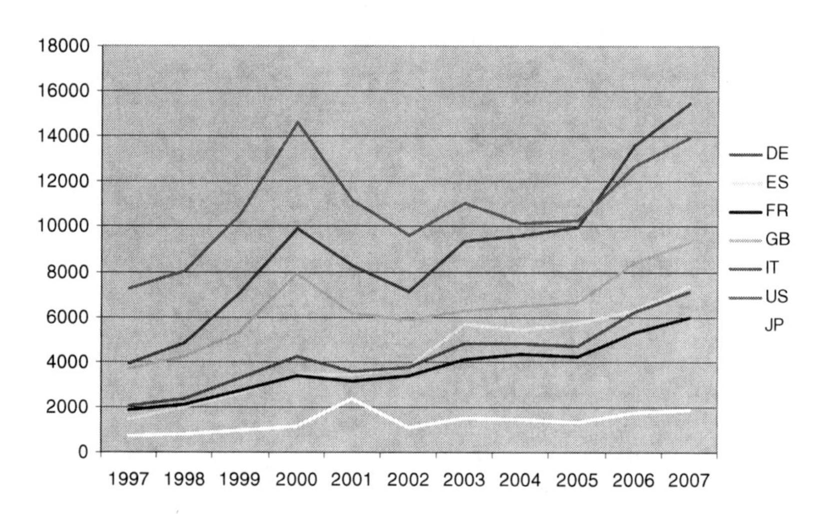

Fig. 2. Applications of CTMs for the largest applicant countries, absolute numbers, 1997-2007[1]

Among non-EU countries US players' interest in the European-wide market is substantially higher than Japanese's, neither of which has yet surpassed their respective 2000-2001 levels where the "New Economy" bubble is clearly visible. Relative positions among the large economies are stable. But here comes a distinctive novelty: Germany has surpassed the US in 2006 as the top user of the CTM system, reaching unprecedented levels in 2007.

[1] OHIM data handed to the author from its central database, own calculations. See [3]. Note: Trends in absolute numbers are reported for large economies generating the highest volumes of applications: Germany (DE), United States (US), Great Britain (GB), Spain (ES), Italy (IT), France (FR), and Japan (JP).

Trademark data also realizes its potential as an innovation indicator when combined with other, more conventional yardsticks of innovative performance. If we go back to the original definition by Schumpeter we realize that is emphasis on innovation as invention plus market introduction can operationalized using a proxy of technological capability (say R&D or patents) and contrast that with the information provided by trademarks. Figure 3 uses country R&D intensity (R&D/GDP) on the x-axis and "trademark intensity" (CTMs/POP) on the y-axis to generate, for the first time, a picture of a possible taxonomy of countries. Although this can be seen as a crude mapping exercise it could be taken as producing some interesting results. We thus can see that there are countries above the average EU-15 both in technological and marketing capabilities (most Nordic countries), as well as those scoring bellow in both (most southern European countries). We can also see that Finland and France appear to be stronger on the technological front, whereas the Netherlands, Ireland and Spain have most of their relative strengths in marketing.

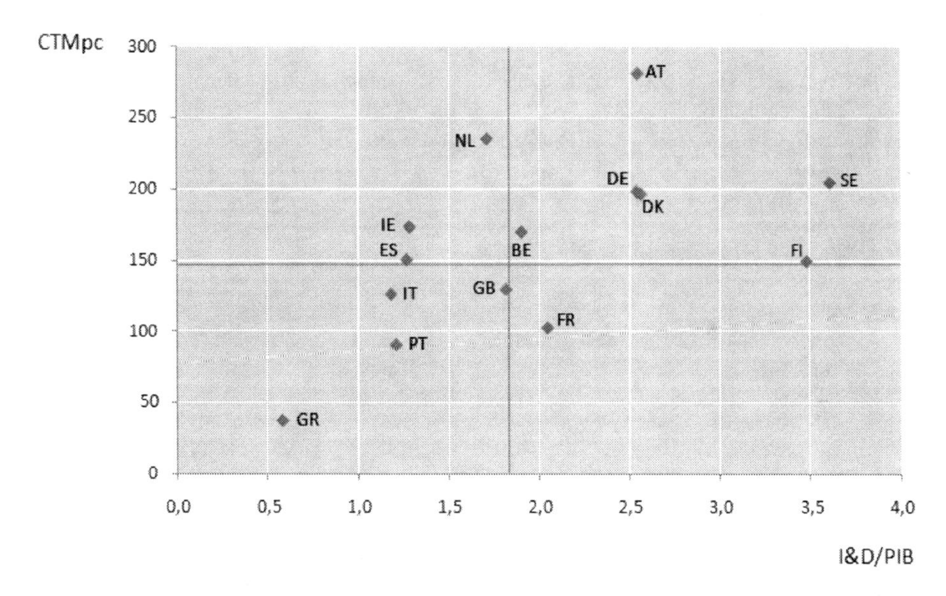

Fig. 3. Technological and marketing capabilities: a country taxonomy [3][2]

5 Countries' Trademark-Intensiveness in Telecommunications

Two classes are of interest as far as telecommunications are concerned, class 9 (containing a whole set of manufacturing goods, including scientific instruments and telecommunications equipment) and class 38 (a service class only referring to telecoms). As Table 1 shows both classes feature in the top 10 most trademarked classes.

[2] OHIM and OECD, own calculations. See [3]. Note. R&D intensity = total expenditures in R&D over GDP; Trademark intensity = CTM application per million inhabitants.

Table 1. Top classes in terms[3]

TOP-10	World		EU-15	
1	Instruments & ICT	10%	Instruments and ICTs	8%
2	Advertising and consultancy	7%	Advertising and consultancy	8%
3	R&D services	7%	R&D services	6%
4	Paper and packaging	6%	Education, culture, sports	6%
5	Education, culture, sports	5%	Paper and packaging	5%
6	Clothing and footwear	5%	Clothing and footwear	5%
7	**Telecoms**	4%	**Telecoms**	3%
8	Pharma and fine chemicals	4%	Pharma and fine chemicals	3%
9	Light chemicals and cosmetics	3%	Light chemicals and cosmetics	3%
10	Finance	3%	Leather and similar materials	3%

Figure 4 isolates the two telecom-related tangible and intangible nice classes (respectively classes 9 and 38). These classes cover what could be broadly described and ICT/Telecom equipment and ICT/Telecom services. The telecoms data shows an upward trend punctuated by large fluctuations. The data reveals spikes around the year 2000 (the "New Economy" bubble) and a stagnation beyond 2007 (the "great recession"/"little depression").

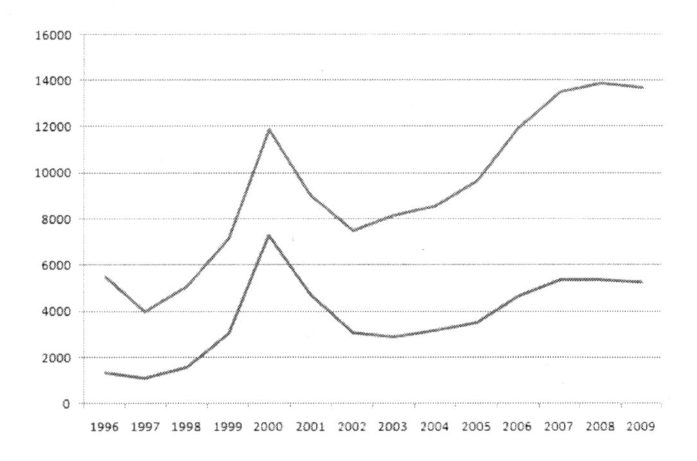

Fig. 4. New CTM applications in telecoms equipment (BLUE) and services (RED), 1997-2009[3]

Figure 5 analyses the concentration by country of new community trademark in telecommunications for the EU-15 in the period 1996-2009. Concentration decreases

[3] OHIM data, own calculations. For the use of trademarks as indicators of service innovation see [3], [4] and [5].

until the early 2000s but that dynamics seemed to have stopped by then. The indicator is the Hirshmann-Herfindahl index [3] and is defined as:

$$IHH = \sum_{i=1}^{k} S_i^2$$

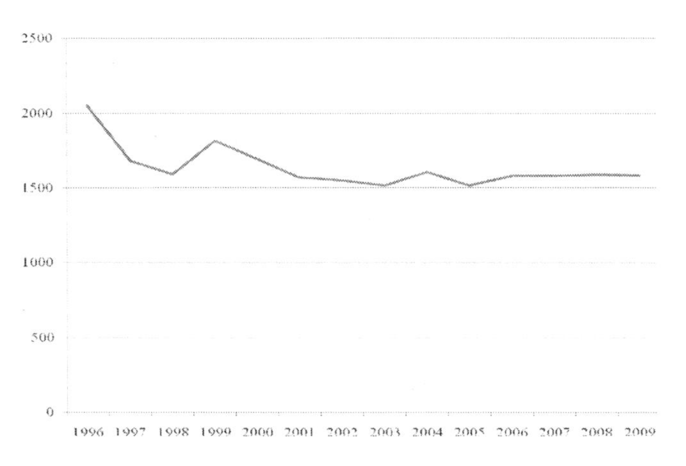

Fig. 5. Concentration in telecom services in the EU-15 (class 38), 1996-2009[4]

The dynamic struggle for market shares is shown in Figure 6, which computes the instability index [3]. It also reinforces the notion that competition as been on the decrease for most of the time. The index is defined as:

$$II = \frac{1}{2} \sum_{i}^{n} \left| S_{i2} - S_{i1} \right|$$

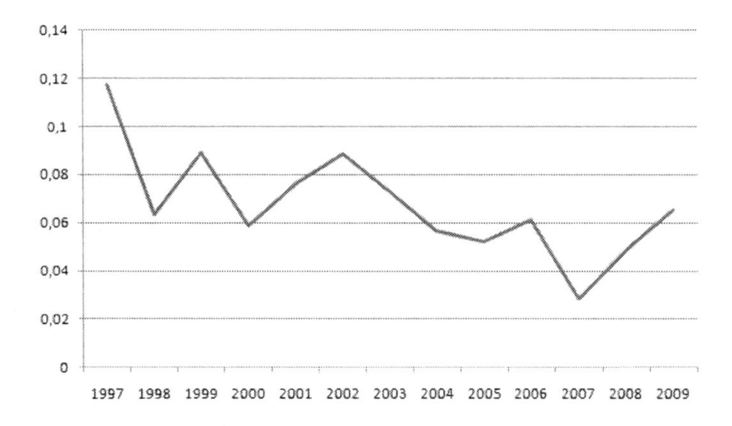

Fig. 6. Market instability in telecom services in the EU (class 38), 1996-2009[4]

---

[4] OHIM data, own calculations. See [3].

Figure 7 computes the "revealed marketing advantage" index for a sample of countries. Is a country is above one it has a competitive strength in a given product class. This approach allows us to put a taxonomy of countries. The index is:

$$RMA_{it} = \left[\frac{tm_{it}}{\sum_t tm_{it}} \right] \Big/ \frac{\sum_i tm_{it}}{\sum_{it} tm_{it}}$$

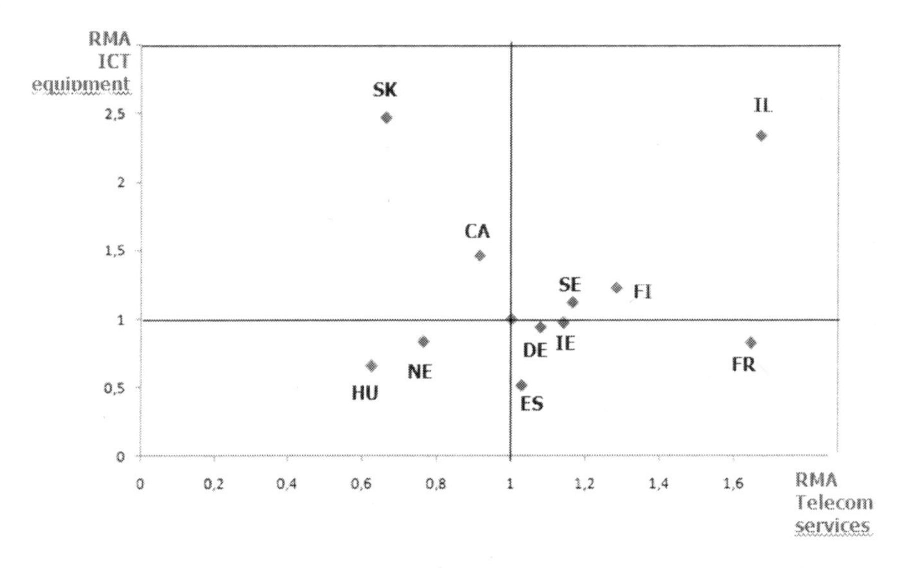

Fig. 7. Revealed marketing advantage in telecoms equipment and services[5]

6 Conclusions

Innovation does not consist exclusively of embodying technological knowledge into high-tech products. Innovation is wider a process that does not stop here; it also implies understanding, segmenting, targeting and persuading increasingly international customer bases of the uses of the new or improved, tangible or intangible artifacts. Thus, innovation is also about establishing the notoriety and credibility of products when introducing them in the marketplace. Typically reliant on technology-oriented indicators such as research and development (R&D) or patents, empirical work on innovation has difficulty in capturing this marketing dimension of innovative activities. The goal of this section is to contribute to fill this gap by staying closely to Schumpeter's definition of innovation. It treats and measures innovation as the outcome of a combination of twin bodies of expertise: technological and marketing. We believe that trademarks are unique but still an under-exploited source of information for studying innovation behavior and industrial dynamics in the telecommunications sector. Thus, this potential should be utilized.

[5] OHIM data, own calculations. See [3].

References

1. Fagerberg, J.: Innovation: A guide to the literature. In: Fagerberg, J., Mowery, D., Nelson, R. (eds.) The Oxford Handbook of Innovation, pp. 1–26. Oxford University Press, Oxford
2. Mendonça, S., Pereira, T.S., Godinho, M.M.: Trademarks as an indicator of innovation and industrial change. Research Policy 33, 1385–1404 (2004)
3. Mendonça, S. (ed.): Estudo sobre o Contributo das Marcas para o Crescimento Económico e para a Competitividade Internacional. INPI (2011)
4. Schmoch, U.: Service marks as novel innovation indicator. Research Evaluation 12, 149–156 (2003)
5. Schmoch, U., Gauch, S.: Service marks as indicators for innovation in knowledge-based services. Research Evaluation 18(4), 323–335 (2009)

Introduction to Social and Policy Implications of Communication Technologies

Antonis M. Hadjiantonis

KIOS Research Center for Intelligent Systems and Networks, University of Cyprus
antonish@ucy.ac.cy

This chapter delves into the far reaching socio-economic drivers and constraints of telecommunications and communications media, and hinges on a diversity of policies not just about technical access to them and pricing, but more importantly about the social and freedom-of-expression rights. After all, communications systems and services are there to serve people's interactions. Therefore this chapter presents four different aspects of telecommunications, dealing with topics largely ignored in the past and in telecommunications and media, i.e. social implications and policy aspects of broadcast media.

Section 1 on "Mobile Communications of the Needy and Poor: Affordability Indicators, European Data, and Social Tariffs" by L.-F. Pau, P. Puga, H. Chen, and Z. Kirtava, is driven by the concerns of the poor and needy, who cannot always afford to be networked, although their hunger for such technology is huge, as inclusion means survival, health, and family linkages. The section is the result of a 4 year data and information collection effort to gather data on the poor and needy in 6 European countries, their mobile communications behaviors, in relation to their socio-economic living contexts. It also provides a new indicator, with case data from 7 countries in 2006 and 2010, allowing to compare the affordability of mobile communications for the poor and needy, by determining the "poor's purchasing power parity (PPP) in wireless minutes per month". Thus, this section provides data showing the social limits of inclusion imposed by the public wireless communications providers, while mobile communications are used today by the vast majority of the world's population.

The work on "Implications of mHealth Service Deployments: A Comparison between Dissimilar European Countries" in Section 2 by F. Vannieuwenborg, Z. Kirtava, L. Lambrinos, J. Van Ooteghem, and S. Verbrugge, discusses the deployment of eHealth and in particular mHealth services for three European countries. Each country uses the same kind of services with a different emphasis. By comparing results of a reference case, tele-monitoring heart patients, the authors identify major implications for a general rollout of the service and formulate recommendations for it.

The work on "Social Communication Behaviors of Virtual Leaders" in Section 3 by D. Shwarts-Asher examines how leadership operates and structures itself inside distant teams linked by communication means. A model, with supporting results from controlled experiments, suggests that communication behaviors and styles of a leader mediate the extent of the virtuality of shared tasks in a team, and thus the team's outputs. A comparison is made with teams working face-to-face. Therefore, this section represents the effect of communications on the way social interaction and organization shape themselves.

A.M. Hadjiantonis and B. Stiller (Eds.): Telecommunication Economics, LNCS 7216, pp. 42–43, 2012.

Finally, Section 4 on "Policy Implications of Digital Television" by B. Sapio is about those factors affecting the usage of digital television-based e-Government services (especially payment services). The focus is on policies driving an adoption by a specific methodology. That set of services is coined under the term "T-government" and pilot study efforts are reported. Comparative data are supplied across several European countries. Thus, this section deals with services enabling social cohesion and citizen's rights, as being possibly distributed by the world's second mostly accessible communications media, which is television as it turns digital.

Mobile Communications of the Needy and Poor: Affordability Indicators, European Data, Social Tariffs

Louis-Francois Pau[1], Pedro Puga[2], Hong Chen[3], and Zviad Kirtava[4]

[1] Copenhagen Business School & Rotterdam School of Management
[2] CIES-IUL Portuguese Center for Research and studies in Sociology
[3] Huawei Technologies Europe
[4] Partners for Health NGO / National Information Learning Centre, Tbilisi, Georgia
lpau@nypost.dk, pedro.puga@iscte.pt, hchen77@gmail.com, zkirtava@nilc.org.ge

Abstract. This section summarizes the data and information collected across Europe and beyond, inside the COST 605 Action (2008-2011), and can be relevant for separate analysis, research and regulations about the mobile communications costs of poor and needy groups. Miscellaneous social and macro-economic data on the needy and their mobile communications usage in five European countries are presented. It also contains the specification of a new indicator whereby the affordability of mobile communications for poor and needy can be established, and the corresponding results for 7 countries in 2006 and 2010. The indicator is the "poor's purchasing power parity (PPP) in wireless minutes per month", based on reported data collected on the distribution amongst poor and needy groups in Georgia and France. An Appendix gives some data sources for the countries where data were collected.

Keywords: e-Inclusion, Mobile communications, Social tariffs, Affordability, Indicators.

1 Mobile Communications of the Needy and Poor

This section summarizes the data and information collected across Europe and beyond, inside the COST 605 Action (2008-2011), and can be relevant for separate analysis, research and regulations about the mobile communications costs of poor and needy groups. The derived analyses and research or regulations are not surveyed in this section, and this section is not structured as a research article but as a fact collection. An Appendix gives some national data sources for the countries where data were collected. A list of references gives pointers to related data collection, methodology or analysis work and also contains the references to the papers produced by the COST 605 Action on communications for the needy [1, 3, 8, 9, 14, 15].

The methodology whereby these data have been collected and aggregated has been to identify and verify relevant, but often inconsistent and unpublished data sources (government or NGO reports, regulator reports, survey reports, UN and ITU statistics, operator data, verbal reports). Mobile tariff and pricing data have used operator data,

A.M. Hadjiantonis and B. Stiller (Eds.): Telecommunication Economics, LNCS 7216, pp. 44–55, 2012.

field work, regulator reports (Europe and Georgia) (Appendix 1), as well as tariff surveys [5, 6]). The reader is thus warned about the fact that, for lack of extensive research budgets and means, the data assembled here could not always be validated from multiple sources, although all of them seem to represent the consensus estimates of operational people working with, or on, poor and needy.

It should be stressed that while there has been extensive data collection in developed as well as developing countries [1, 4] on demographics, revenues , housing of poor and needy, there has been almost none on the telecommunications practices and spending of these groups, or on specific telecommunications services (even on public telephony). There has been theoretical analysis rooted in income distribution [11] the applicability of which hedges on data collection efforts like this one. This section furthermore focuses on public wireless communications, which represent by far the widest relative adoption in that group among combined fixed telephony, broadband, Internet and wireless services [13]. In terms of geographical coverage, the focus is on the much neglected European poor and needy, as well as those in Georgia, representing a population on European fringes and part of the Council of Europe [10]. Whereas OECD has defined baskets of telecommunications services with variants in 2002, 2006 and 2010, the poor /needy groups for basic practical reasons and affordability still prioritize plain voice communications the highest for social inclusion, work, and other specific needs. While this fact may evolve over time, focus is here solely on wireless voice and the corresponding usage in view of the affordability constraints of these groups.

World Development Indicators Database (World Bank group) methodology for mobile pricing is using a monthly basket of 25 "standard" calls + 30 SMS. One of OECD's baskets includes per year: 1680 outgoing calls (2952 min) + 600 SMS + 12 MMS. ITU's ICT Price Basket (IPB) index and its depiction of ICT affordability is a composite affordability measure based on three sub-baskets – fixed telephone, mobile cellular and fixed broadband Internet services – and computed as a percentage of average Gross National Income (GNI) per capita. ICT spending correspond to less than 5% of GNI per capita in much of Europe and Asia and the Pacific's high income economies, as well as in the US and Canada; but for sure, for poor and needy, which have chosen to be connected (prepaid or post-paid), the ratio is much larger, making the issue much more critical. The ITU IPB studies also reveal a close link between the affordability of ICT services and national income levels: people in high-income countries spend relatively little for ICT services, while those in the world's poorest countries spend relatively more, both in absolute terms (e.g., mobile minute cost) and in relative terms (as a share of disposable income).

2 Who Are the Poor and Needy

This section aggregates, or segments, data about the following groups in the population [13]:

- Poor, defined as such under minimum household subsistence revenue limit [4]
- Homeless (still approx. 30 % have work)
- Unemployed getting unemployment benefits, searching jobs , but under minimum household subsistence limit

- Unemployed after expiration of unemployment and social benefits
- Isolated individuals of all ages, often subject to a social / medical / penal fracture (e.g., 39 % of age group 79-83 years in France)
- Migrant workers who have or find very short term employment away from domicile (incl. seasonal workers in, e.g., tourism and agriculture) [2]
- Elderly alone with home care on low pensions under subsistence limit (e.g. 89 % of 79-83 years age group in France)
- Displaced populations due to war, national disasters or climate change effects
- Some immigrants (approx. 10 million immigrant workers in Europe) [2, 5]
- Many categories of disabled persons and some categories of sick persons

According to Eurostat this heterogeneous group represented in 2004: 73 Million people in EU (16 % of the population), 2 Million people in EFTA countries, and 160 Million people in the geographical areas part of Council of Europe [7]. In the above segmentation, the household subsistence revenue limit is usually defined as by UN at 60% of national GNI average.

2.1 European Macrodata on the "Needy" and Their Mobile Communications Usage or "Social Mobile Tariffs"

Some COST 605 Action participating countries have contributed to a survey on the above subject (Denmark, France, Georgia, Netherlands, and Portugal). The focus was on needy / poor people (see subsection 2.1), especially on

- social parameters , demographics, employment, income,
- measured or assumed mobile communications service usage and spending, and
- cases of individuals or groups of needy people's as to mobile communications usage and spending (from individual interviews, secondary information sources).

Table 1 addresses data collected so far across 5 EU countries mostly for 2006: Netherlands, Portugal, France, United Kingdom, and Denmark. It should be noted that definitions are not always identical and that collection methods and sources may be inconsistent. It should also be noted that in general the data are from many different sources thus not always with same year or basis. Some data were collected on the basis of field surveys and are not found in other sources. The Monetary unit was Euros for calibration purposes (except UK where GPB/EUR exchange rate fluctuations prevented that calibration). When a definition cannot be used uniformly across countries, due to different legislations, or data collection methods, indications are given in the Table as to different interpretations; especially UK has different estimation concepts (see Appendix).

The goal of the survey is to provide the basis for the establishment of social wireless tariff bundles made affordable to the needy and subsidized or not by operators and/or regulators (using revenue from universal service obligation laws, where applicable) [5, 10]. Another goal is to help European regulator and the Commission prepare for directives with the same goal across the EU [5, 6, 7]. As the ubiquity of wireless communications has a high social value and impact, this study is justified beyond general analyses of universal access to broadband services.

Table 1. European macro-data on the "needy" and their mobile communications usage

2006 DATA	NETHERLANDS	PORTUGAL	FRANCE	UK	DENMARK
1.National official poverty threshold per household (from national Social laws)	875 Euro/month or 10500 Euros per year ; "minimum necessary income" of 1000 Euros/month (2005); set by authorities	385 Euro/month; set by authorities	2006 value: 817 Euro/month (determined as 60 % of median disposable income) ; 2008 value : 880 Euro/month 2009 update: 954 Euros/month (determined as 60 % of median of 19080 Euros/year)	around £ 1180 / month (60% of median); in UK this is per equalised household (i.e. adjusted to take account of household size and composition)	
2. Number of residents in households under national poverty threshold (defined above)	1 435 000 (8.9%)	1 907 837 (after social transfers) ; about 18 % of population	1996 value: 4,5 Million (including 950 000 children aged 14 or less living in 1,8 M households) 2006 value: 7 136 000 (13,0 % of population) 2008 value : 7,8 Million or 13,2 % of population (of which more than 900 000 retired) (INSEE) 2009 value: 8,17 % of population or 13,5 % of population (estimate of 4,5 M under threshold of 773 Euros) (INSEE)	Around 17% of households or 10 Million people; 9.2 Million people are below the poverty line measured before, and 11.4 Million after, housing costs	6 % of population
3. National median revenue per household		643 Euro/month	1362 Euro/month ; 2008 value: 1470 Euro/month (INSEE)	£ 1968 / month	
4. Minimal social insertion income (public grant to poorest eligible working age individuals who accept it)	780 Euro/month (AOW); set by authorities	479 Euro/month; set by authorities	350 Euro / month; set by authorities	£ 192 / month for single people under 25, £ 240 / month a week for single people 25 and over	
5. Number of individuals receiving social insertion income (stated above)	657 000 (2004)	264 287	1 180 000	5.3 Million people of working age receive income-related benefits	
6. Number of pension age individual recipients of a public grant (typically when pension is under poverty threshold 1.; grant is a top-up or replacement for pension)	122 000 (2004)	671 047	1 080 000	3.3 Million pensioners receive pension credit ("top up" to state pension)	
7. Number of prison inmates (for over 1 month)	16 000	12 637	61 700	78 000	
8. Number of youngsters (<16 y) aged less than working age, in households with at least one adult under minimal social insertion income level (defined in 4.)	461 000 (14 %) (2004)	350 661 (20 %)	4 150 000	2.9 Million children living in households below the poverty line (22% of children)	

Table 1. *(Continued)*

2006 DATA	NETHERLANDS	PORTUGAL	FRANCE	UK	DENMARK
9. Number of homeless sleeping in streets, etc. (estimates, averaged over year)			30 000		5000 (2007 and 2009 data) of which 3000 in Copenhagen; 69 % had dependency problems (alcohol, drugs, medicine) ;75 % received cash assistance
10.Mobile penetration in group (2) and median individual ARPU spend/month		Total penetration rate: 115 %	37 % / 35 Euro/Month	Using lowest quintile as a proxy for group (2), 59%, combined spend on fixed and mobile £ 25,60 / month	
11.Mobile penetration in group (5) and median individual ARPU spend/month			14 % / 10 Euro/Month		
12. Mobile penetration in group (6) and mean individual ARPU spend/month			9 % / 21 Euro/Month	38% mobile penetration for single retired people and 56% for retired couples	
13. Mobile penetration in group (8) and median individual ARPU spend/month			56 % / 29 Euro/Month		
14. Avg. spend on TV equipment amortization and media subsc.in group (2)			17 Euro/Month/household	£ 13 / month spent by lowest quintile on TV and internet together	
15. Average spend on Internet access and content in group (2)			6 Euro/Month/household	£ 13 / month spent by lowest quintile on TV and internet together	
16. Range on 3 room 60 m2 rental cost averages in cities above 100 000 inhabitants	Avg. Housing (22%) and Avg. water-energy (8%) of 860 Euros/month		574-1456 Euro/Month		
17. Minimum regulated price of one consultation at doctor		2,15 Euro (family doctor)	22 Euro	Free; regulated revenue per GP per patient approx. 40 £	
18. Social legislation subsidizing directly needy' s communications needs under specific conditions	No	No	Yes (10 Euro/month) ;See Note 1 below.	Not really.	
19. Regulations for Universal service provisioning by operators but with no user subsidies ;	Fixed : Yes Mobile : No Internet : No	Fixed : No Mobile : No Internet : No	Fixed: Yes Mobile: No Internet : No	Fixed: Yes Mobile: No Internet : No	Fixed : No Mobile : No Internet : No
20. Personal insolvency rate in total population for non-business reasons (Court decisions)	30 % of persons with low income had negative net assets ; 14 % had non judicial delays in payment of loans; 15 000 legal insolvencies	905 processes	Approx. 250 000 processes (0,15 % of working population)		

Table 2. 2006 Indicators of a poor's purchasing power parity (PPP) in wireless minutes/ month; the currencies used are those attached to the source data collected; exchange rates are per end 2006

RANK	COUNTRY	**POVERTY INDICATORS 2006** Gross National product/ inhabitant on Purchasing Power Parity basis (IMF www.undata.org ,2006)	**MOBILE TARIFF 2006** www.erg.eu.int ,Tarifica , Intelecon (prepaid, no discount, average, national call)	**POOR's 2006 PPP MOBILE MINUTES/month** (2006 exchange rates; 1 $= approx. 1 Euro)
1	France	GDP PPP 33509 $/y -7,9 M poor (INSEE, or 13,6 % of pop.) (< 882 Euro/mo., 2006) ,highest in towns of > 20 000 -100 000 homeless (Source: Fondation Abbé Pierre)	ERG average 0,0980 Euro/min -550 000 households have no possible Internet access	17 096
2	Denmark	GDP PPP 36700 $/y	ERG average 0,136 Euro/min	11 059
3	Portugal	GDP PPP 21779 $/y	ERG average 0,1319 Euro/min	8 255
4	Mexico	GDP PPP 14120 $/y -2nd Poorest quarter 92 $/mo (30,9 % pop) -Poorest quarter 32 $/mo (34,6 % pop)	0,49 $/min -Entry barrier 32 $ + min. recharge 9 $/instance	1 440
5	Georgia	GDP PPP 4400 $/y (GDP 1870 Euros) -22 $/month pensions	0,165 $/min (0,28 GEL/min)	1 333
6	Morocco	GDP PPP 2400 $/y - 21 % of pop under poverty line	0,14 $/min - Average - Prepaid 96 %	857
7	Tanzania	GDP PPP 1256 $/y	0,31 $/min -50 USD handset = 1 year of savings	202

Table 3. 2010 Indicators of a poor's purchasing power parity (PPP) in wireless minutes/ month; the currencies used are those attached to the source data collected; exchange rates are per end 2010 ; (*): prepaid, no discount, average, national call

RANK	COUNTRY	**POVERTY INDICATORS 2010** Gross National product/.inhabitant on Purchasing Power Parity (http://data.worldbank.org/indicator/ NY.GNP.PCAP.PP.CD)	**MOBILE TARIFF 2010** http://www.erg.eu.int/ (BEREC) , Tarifica , Intelecon , Numbeo.com, lowcostmobile.fr (prepaid, no discount, average, national call)	**POOR's 2010 PPP MOBILE MINUTES/month** (2010 exchange rates)
1	Denmark	GNI PPP 40290 $/y	0,15 $/min	13 430
2	France	GNI PPP 34440 $/y	0,32 $/min	5 381
3	Portugal	GNI PPP 24710 $/y	0,25 $/min	4 942
4	Mexico	GNI PPP 14360 $/y	0,31 $/min	2 316
5	Georgia	GNI PPP 4960 $/y	0,145 $/min (0,257 GEL/min)	1 710
6	Morocco	GNI PPP 4620 $/y	0,67 $ /min	344
7	Tanzania	GNI PPP 1420 $/y	0,35 $/min	202

2.2 The Needy's Purchasing Power in Mobile Minutes and the "Mobile Communications Divide"

As an attempt to compare by a single indicator the affordability of mobile communications for poor and needy groups across countries [13, 14], we present the following indicator: "A poor's purchasing power parity (PPP) in wireless minutes per month". PPP is defined as: the ratio of UN purchasing power GNI PPP (Gross National product per Inhabitant on Purchasing power parity) (in one basis currency), multiplied by 0.6 to reflect the normal definition of the poverty limit, divided by the average price (in same basis currency) of one wireless voice minute (fully loaded to caller) as established by the national regulator, and further divided by 12 months. This wireless voice minute is assumed prepaid, with no terminals cost share, for a call to a national number, excluding discounts and plans. It shows, in simple words, how many minutes of wireless voice a poor would get per month if all his purchasing income that month was spent prepaid on talking in the same country over a wireless operator.

Table 2, gives 2006 data for 7 countries of which three EU countries addressed in detail in subsection 2.2, and it was elaborated for the first time by the authors in Lisbon in 2008. Table 3 below gives an update for 2010. These tables show:

- For each of them, very significant *ranges* of values: in 2006 from 202 minutes/month (Tanzania) to 284 hours/month (France), and in 2010 from 202 minutes/month to 223 hours/month (Denmark); this represents the "mobile communications divide";
- Fairly stable *rankings* in the small country sample, but that the index for some countries has fallen significantly (France, Morocco, Portugal), while it has risen for some others (Mexico, Denmark);
- The average mobile tariff *dynamics* over time do not correlate with the evolution of the PPP GNI (Gross National product/inhabitant on purchasing power parity basis). This is despite regulatory interventions, competition, reductions in interconnection rates (European Union), within-operator's network vs. outgoing call differentiation (Europe), specific tariff plans or brands for social groups (for "unemployed" like in Spain, for "poor and needy" like in France, "youth" everywhere), and/or occasional ceilings on price per mobile minute (as in Georgia: 0,24 GEL/min since 2011).

Some consultancies adopt non-transparent rank determination methods, like the case of INFORMA Telecoms & Media and World Development Business "awarding" 3rd place to Georgia (among 186 countries researched) for having one of the most expensive mobile rates in the world (Jan 2010). Most consultancies studying e-Inclusion just do not include wireless access, but only Internet access [5].

Example of "Social Mobile tariff" (France): A France Telecom / Orange offer called "RSA ("Revenu de solidarité active") special tariff plan", available in Metropolitan France, and linked to RSA legislation and recognition. It includes 40 minutes of calls to landlines and mobiles plus 40 SMS, for €10 per month, without a contract. It is available to all RSA beneficiaries, whether or not they are already Orange customers. This plan allows unused minutes and SMS to be carried forward.

Orange customers eligible for the "RSA special tariff plan" but already under contract can move onto this plan at no cost and with no commitment. Customers from other operators can keep their existing mobile number. Orange is also selling a range of cheap mobile phones in its stores starting from €39. RSA beneficiaries only can either buy second-hand handsets from €10 or benefit from the offer without buying a mobile" Orange already offers several benefits to promote spending power in France:

- unlimited SMS service or 10% reduction on some offers for young people under 26 years old;
- 10% reduction on "initial" fixed rate packages for those over 60 years old;
- 20% reduction on the "click" offer for jobseekers and large families.

Furthermore, a €20 social triple play offer (telephone, TV and broadband Internet) is being prepared

2.3 Distribution by Groups of the Poor and Needy and Effects of Falling Affordability

This subsection provides the distribution of poor and needy for Georgia and France by narrower groups as those identified in subsection 2.1 and expands this analysis for household's general affordability constraints in Denmark. Table 4 gives for France the distribution by categories of the poor under the poverty level; it also gives the percentage in each category with e.g. for 2009 a maximum of 34,7 % for unemployed and 30,3 % for "other inactives". Table 5 contains the distribution for Georgia for an estimated total population of needy and poor of 4, 4 Million inhabitants. The Danish data [12] confirm falling affordability due to falling disposable private consumption and fast rise in tariffs and address the structural analysis of affordability data for households. The growing overexposure of households towards real estate and consumer credit, has lead the credit ratio (balance of liabilities/ net disposable income after taxes) to raise to 300 % , but assets (incl. Pension assets and free shares in real estate) have grown faster with a ratio of 460 % (2010). Inside liabilities, real estate mortgage debt represents about 75 % and other liabilities 25%. The combination of high liabilities, illiquid assets and falling real estate prices, has increased the risk from loss of income/job and has exposed the households to interest rate increases; a 3 % increase in interest rates at the end of 2010 lead to interest payments reaching avg. 30% of net income. Due to parallel increases in energy and transportation costs, the private consumption ratio (disposable income net of taxes and energy) in 2010 was only 92 % of the 2007 value.

As median mobile tariffs in Denmark have over the past 10 years grown at approx. twice the inflation rate (Q3:2011: 3%), and as a 2 year postpaid subscription plans can be assimilated to a 2 year mortgage obligation, their share inside net disposable income has approx. doubled; unless short term unsecured consumer loan credits are used to cover this increase, the research question is which expense items had to be reduced to accommodate mobile communications inside falling private consumption.

Table 4. Distribution of the poor and needy by groups, for metropolitan France (2008) (Source INSEE 30/8/2011)

2008	Thousands of poor persons	Poverty ratio (%)	Poverty intensity (%)
Active persons 18 Years or above	2635	9,5 %	20,6 %
Active in a job, of which	1863	7,3 %	18,2 %
-Salaried	1445	6,3 %	15,8 %
-Independent workers	418	15,3 %	29,1 %
Unemployed	772	35,8 %	27,2 %
Inactive aged 18 y or above	2873	15,1 %	17,2 %
-Students	324	18,1 %	19,1 %
-Retired persons	1283	9,9 %	13,0 %
-Other inactive	1266	29,3 %	21,4 %
Children < 18 years	2328	17,3 %	18,3 %
Total population	7836	13,0 %	18,5 %

Table 5. Distribution of poor and needy in Georgia (2008); Source: National Statistics office of Georgia www.geostat.ge

GEORGIA 2010 Categories	% of total needy	Inhabitants
Refugees/IDP's	6.1%	270000
Unemployed	16.4%	315000
Pensioners	14.9%	656000
Disabled	3.1%	137000
Impoverished	8.4%	370000
Near poverty level	14.3%	630000
Prisoners	0.4%	18659
Unspecified	45.4%	1997000
TOTAL	100 %	4 394 000

3 Summary and Conclusions

The present data analysis in several countries provides key data to analyze the affordability of mobile communications for the poor and needy. It also presents several simple indicators to be used for policy making and tariffing. Mobile social tariffs are also described to reduce the mobile communications divide, which is at least as important as the Internet digital divide.

References

1. Avila, A.: Underdeveloped ICT areas in Sub-Saharan Africa. Informatica Economică Journal 13(2) (2009)
2. Calenda, D. (ed.): Migrants, ethnic minorities and ICT IN Europe - National Scenario: an overview of six countries (Germany, Spain, Italy, France, The Czech Republic, The Netherlands), BRIDGE-IT Thematic network (2011)
3. Concepción Garcia-Jimenez, M., Gomez-Barroso, J.-L.: Universal Service in a Broader Perspective: The European Digital Divide. Informatica Economică Journal 13(2) (2009), `http://www.revistaie.ase.ro/current.html`
4. Dilip Potnis, D.: Cell-phone-enabled empowerment of women earning less than $1/day. IEEE Technology and Society Magazine, 39–45 (Summer 2011)
5. Empirica (for EU), Benchmarking in a Policy Perspective: Report no 5 on eInclusion. Gesellschaft für Kommunikations- und Technologie-forschung mBH, Bonn (2006)
6. European Commission, Report on Telecoms Price Developments from 1998 to 2010 (2011)
7. European Commission, Digital agenda assembly, 16-17/6/2011, Report from workshop 2009 Access and digital ability: building a barrier-free digital society
8. Falch, M., Henten, A.: Achieving Universal Access to Broadband. Informatica Economică Journal 13(2) (2009)
9. Kirtava, Z.: Mobile Operator Supports Refugees and Disabled in Georgia. Informatica Economică Journal 13(2) (2009)
10. Kirtava, Z.: Mobile Operators For Low-Income Population - Case of Georgia: Special Tariffs - To Be or Not To Be? In: Proceedings of COST 298 Project Conference on Broadband Society "The Good, The Bad and The Challenging – The User and the Future of Information and Communication Technologies", Copenhagen, May 13-15, vol. II, pp. 930–939 (2009)
11. Milne, C.: Affordability of basic telephone service: an income distribution approach. Telecommunications Policy 24, 907–927 (2000)
12. Nationalbanken, Finansiel stabilitet, pp. 70–78 and Kvartalsoversigt, Q2-Del 1, pp. 39–44 (2011) (in Danish)
13. Pau, L.-F.: Mobile service Affordability for the Needy, Addiction, and ICT policy implications. In: Proc. 7th Intl. Conf. on Mobile Business, Barcelona, July 7-8 (2008)
14. Pau, L.-F.: Enabling mobile communications for the needy: affordability methodology, and approaches to re-qualify universal service measures. Informatica Economică Journal 13(2) (2009)
15. Puga, P., Cardoso, G., Espanha, R., Mendonça, S.: Telecommunications for the Needy: How needed are they? Informatica Economică Journal 13(2) (2009)

Appendix: Data Sources

1. **Europe**: BERC (European regulator's group) www.erg.eu.int
2. **United Nations**: UN data www.undata.org
3. **Denmark**: Nationalt forskningscenter for velfaerd www.sfi.dk
4. **France**: French regulator ARCEP www.arcep.fr; Fondation Abbé Pierre http://www.fondation-abbe-pierre.fr/; Fondation de France http://www.fdf.org/; Petits frères des pauvres http://www.petitsfreres.asso.fr; INSEE www.insee.fr, Observatoire national de la pauvreté et de l'exclusion sociale (ONPES) http://www.travail-solidarite.gouv.fr/web/observatoire-national-pauvrete-exclusion-sociale; INSEE study «Revenus fixes et sociaux 2006-2009, 30/8/2011; Ministère de l'emploi et de la solidarité http://www.interieur.gouv.fr/misill/sections/liens/ministere-emploi-solidarite/view France Telecom social tariff: http://www.francetelecom.com%2fen_EN%2fpress%2fpress_releases%2fcp090512en.jsp; Handicap International http://www.handicap-international.fr/; French Parliament decision on social tariffs of 09 June 2008 : http://www.assemblee-nationale.fr/13/cra/2007-2008/187.asp
5. **Georgia**: Georgia National Communication Commission (GNCC) Annual reports 2005-2010 (in Georgian) www.gncc.ge ; Georgia Social Service Agency (in 2008: Social Subsidies Agency) database http://ssa.gov.ge ; National Statistics Office of Georgia www.geostat.ge
6. **Netherlands**: National Bureau of statistics www.cbs.nl/en-GB/
7. **Portugal**: DGERT (Direcção-Geral do Emprego e das Relações de Trabalho - Ministry of Solidarity and Social Security): www.dgert.mtss.gov.pt/ The Portuguese Social Security System: www2.seg-social.pt; INE – Statistics Portugal: www.ine.pt Deco - Portuguese Association for the Consumer Defense: www.deco.proteste.pt; ICP- ANACOM: www.anacom.pt
8. **Spain**: Telefonica social tariff: http://www.telefonica.es%2fon%2fio%2fes%2fteayudamos%2fhome.html
9. **UK** : National poverty http://www.statistics.gov.uk/downloads/theme_social/Family_Spending_2007/FamilySpending2008_web.pdf and http://www.dwp.gov.uk/asd/frs/2003_04/tables/pdf/3_8.pdf; Social insertion income www.dwp.gov.uk; Prisoners http://www.homeoffice.gov.uk/rds/pdfs06/hosb1106.pdf; Expected spending on mobile phones for different household types http://www.minimumincomestandard.org; Attitudinal research by Ofcom among low income groups http://ofcom.org.uk/research/tce/ce07/annex4.pdf; Ofcom research into mobile take-up and spend http://ofcom.org.uk/research/tce/ce07/research07.pdf; Affordability report 2006 http://www.regulateonline.org

Implications of mHealth Service Deployments: A Comparison between Dissimilar European Countries

Frederic Vannieuwenborg[1], Zviad Kirtava[2], Lambros Lambrinos[3], Jan Van Ooteghem[1], and Sofie Verbrugge[1]

[1] Ghent University, IBBT, Dept. of Information Technology (INTEC)
[2] Partners for Health NGO, National Information Learning Centre
[3] Cyprus University of Technology, Dept. of Communication and Internet Studies
{frederic.vannieuwenborg,jan.vanooteghem, sofie.verbrugge}@intec.ugent.be, zkirtava@nilc.org.ge, lambros.lambrinos@cut.ac.cy

Abstract. Despite several successful pilot studies, a general implementation and adoption of mobile health services is not often the case. The deployment of eHealth and in particular mHealth services is discussed for three non similar European countries: Cyprus, Georgia, and Belgium. Each country uses the same kind of mobile health services with a different emphasis. By comparing results of a reference case, tele-monitoring heart patients, major implications are identified for a general rollout of services and respective recommendations.

Keywords: mHealth, Tele-monitoring, Business model implications.

1 Introduction

Driven by an aging population, the European demand for care and cure will strongly increase. Also the growth of medical requirements in remote locations in developing states, lower availability of professional care providers, and all this within a stringent health care budget will force us to change the actual way of providing good quality care. Next to revaluation of the care sector and optimization of the work efficiency, integration of ICT in the care sector already has proven its positive impact.

eHealth, and in particular, one of its service – telemedicine; diagnosis and treatments at a distance – currently finds application in the majority of medical domains: radiology, cardiology, dermatology, psychiatry, dentistry and paediatrics. Telemedicine, offering quality services with significant cost savings, has given a new dimension to the healthcare sector. Various telemedicine programs have proven to be cost effective [3] [23] [24], and with the curtailment of health care expenditures, programs as such are being implemented in several nations [7].

2 eHealth: Actual Implementation

Two sorts of services should be distinguished. The most obvious mHealth services are those that are regulated. This type of services is developed within a set of regulations

A.M. Hadjiantonis and B. Stiller (Eds.): Telecommunication Economics, LNCS 7216, pp. 56–66, 2012.

(e.g., privacy or data presentation) to support the professional care providers with professional data. In contrast, non regulated health services are rising under the form of numerous web- and smart phone applications. They can also provide useful information for the caregivers, but are much less restricted by legal facets.

2.1 Regulated Services

eHealth encompasses a range of services or systems that are at the edge of medicine/healthcare and information technology, including (but not limited to):

- Health Management Information Systems (HMIS), Electronic Health/Medical Records (EHR/EMR) and Electronic Health Data Registries,
- Telemedicine: diagnosis and treatments at a distance,
- Health professionals' distance learning and re-training, and
- Health knowledge management through medical informatics resources: *e.g.*, Evidence Based Medicine (EBM) or best practice guidelines on Consumer health/personal informatics, health literacy: use of health/medical e-resources by healthy individuals or patients.

eHealth is seen as recovery and growth tool for healthcare sector by leading economies in the world. The US government has presented a new Strategy for American Innovation on Feb 4, 2011, where Health IT products are named as one of the most important driving forces for improving cost-efficiency and driving US economy towards sustainable growth. According to IDC Research (2008) Western EU market for eHealth will grow from $9 billion (in 2006) to $12 billion (in 2011). A similar trend of doubling – from $16 billion to $35 billion – is foreseen for the US Health IT market. Other continents like Asia and Europe also stimulate heavily the implementation of eHealth applications in order to curtail governmental expenditure on healthcare services [21].

In developing economies or countries with many widespread inhabitants, eHealth applications (e.g. telemedicine, tele-learning or tele-radiology) play an important role to bridge the gap of medicine between rural and urban areas. Typically medical facilities in rural domains do not always offer the same quality level as urban healthcare centers [7].

Mobile Telemedicine (Mobile Health, M-Health) represents a rapidly growing branch of telemedicine services and more specific, providing continuous medical care (at-home and outdoors) preferably based on wireless sensors and actuators communicating with different health actors (hospital specialist, General Practitioner, Medical Insurance Company) via GPRS, 3G or LTE connections. Thanks to the growing use of mobile phones, increase in network coverage, bandwidth rich applications, technical improvements, financial affordability of broadband applications and the lower tariffs for data transmission, mobile telemedicine is evolving as an advantageous and cost-effective tool. The overall objective of mobile telemedicine is the development and trial of new services and applications in the area of mobile health, promoting the use and deployment of GPRS, 3G or LTE mobile services and technologies. The target is to provide the means contributing to the reduction of costs in hospitals and health care, by allowing patients to have complete and personalized monitoring of their health in-home, while pursuing a normal life (instead of being confined in hospital for long periods of monitoring).

Most efficient mHealth applications regard remote monitoring of mainly outpatients with different chronic conditions which cannot be treated at wards continuously but would benefit by having extended remote monitoring:

- Diabetes patients monitoring
- Patients with risk of blood clotting (myocardial infarction, transient ischemic attack /stroke, deep veins thrombosis, pulmonary artery thromboembolism)
- Bronchial Asthma patients
- Arterial hypertension
- Cardiac Arrhythmias
- Epilepsy
- Chronic Heart Failure

2.2 Non-regulated Services

Next to all regulated forms of eHealth and mHealth applications (approved and adopted by (un)professional caregivers and regulators) also unregulated applications, under the form of Internet- and smart phone applications, are expected to gain importance in the care sector. With the growing use of smart phones and all kind of sensor devices, the number of personal informatics services, which are applications allowing people to track their own habits, movements and thoughts, grows exponentially [14] [17]. According to the Global Mobile Health Market Report 2010-2015 [15], in 2015 more than 500 million people will have a mHealth application on their smart phone. The resulting data of these activities such as your personal sleeping pattern, your heartbeat during an exercise, your daily weight, will be an unregulated, but useful completion for a personal health record [15]. Some cardiologists are already tracking patients conditions based on data provided by a Polar® application.

3 Reference Case: Heart Failure Monitoring

Because there already exist numerous projects [4] [20] [2] performed all over the world, which monitored patients with heart failures, we choose to use this mobile application as reference case. This pool of available data allows us to form a good image of the operational processes, the financial and juristic characteristics of mobile monitoring of heart failure patients.

With the actual available technology, mobile monitoring of heart failure patients is relatively easy to do. Generally the infrastructure consists of a central server and/or a call center (Figure 1), a mobile phone and medical devices (an Electrocardiogram (ECG) device, a weighing scale and a monitor for the blood pressure) suited for monitoring arrhythmia/heart failures. A patient can monitor his/her vital body parameters regularly from anywhere (e.g., home or office) and the data is transferred to the server via the mobile phone or gateway. The server analyses the data for any 'red alerts' or 'technical alerts'; also readings are stored in the patients' personal electronic records in the server. In case of any emergency, an alert is sent to the GP, nurse and/or cardiologist, who can access the health records via mobile phone or computer connected to Internet, and correctly diagnose, prescribe additional medicine, take action.

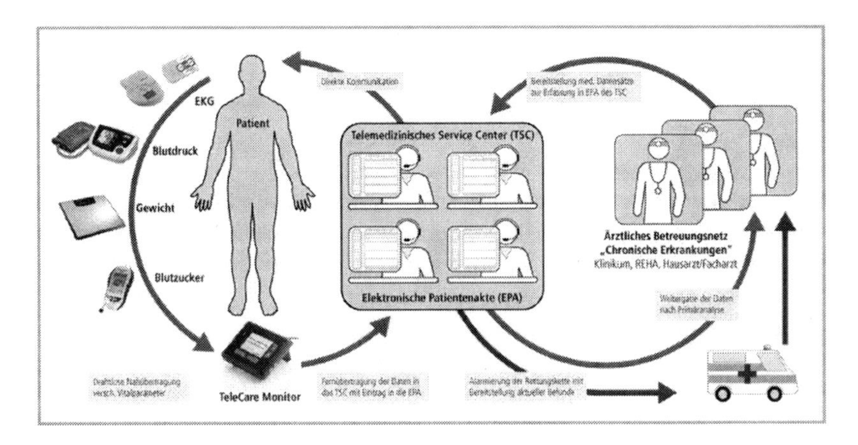

Fig. 1. Tele-monitoring patients with heart failures, source [6]

Tele-monitoring creates a win-win situation for the patient, doctor, hospital and insurance provider [20]. A patient is reassured that his/her health is in good hands, by being silently monitored all the time. Together with better disease education and acceptation, what grows noticeable, this leads to immediate improvement of the quality of life for the patient. Doctors can pursue other activities or manage different patients. At the same time they still can be reached easily anywhere and anytime. Also, historical data can reveal insights in the development of the disease. This could be a valuable input for future research and development in the pharmaceutical and medical field. Hospitals can discharge patients earlier, and insurance companies do save funds with reduced number of hospitalizations and shorter in-hospital (ER or specialized department) stays [4] [6] [1].

4 Cases

In what follows we describe a general overview of the *'as is'* state in Georgia, Belgium and in Cyprus. By doing so, we hope to identify some of the implications for a generalized rollout of this telemedicine application.

4.1 Actual Status: Georgia

Georgia is a developing low and middle income country that recently reformed a part of its healthcare delivery system. It changed from a universal low-cost healthcare model that was inherited from the Soviet Union, to a largely fee-for-service system.

Cardiovascular diseases, especially Arterial Hypertension (AH) and Ischemic Heart Disease (IHD) are the main cause of both morbidity (mainly AH) and especially mortality (mainly IHD) in Georgia.

According to statistics provided by World Health Organization [25], in 2002 Georgia was among the five leading countries (Bahamas, Georgia, Lesotho, Liberia and Maldives) with the highest incidence of AH. Based on the WHO Surveillance of Chronic Risk Factor Report [28] for 2005, adults from Georgia had the second highest average blood pressure in the world. According to most recent statistics provided by Georgian National Center for Disease Control and Public Health, the incidence of AH in Georgia reached 55% by the end of 2009 exceeding the alarming incidence of high blood pressure (43%) as found for African Americans [27].

One of the major complications of both AH and chronic IHD is heart failure. In order to anticipate these risks, they imported a heart failure disease management program (HFMDP) to better follow up and start monitoring their chronic heart failure patients. Results of this implementation are promising less emergency department visits and hospitalizations. On the other hand, more Beta-blockers are prescribed [8].

Finally, one of the leading causes of sudden death and patient disability among cardiology patients are cardiac arrhythmias. Regretfully, until now, Georgian health statistics do not register separately numbers of arrhythmias. However, by approximation of European data we can estimate that there are around 70,000 patients with arrhythmia annually in Georgia. Those patients often need ambulatory monitoring and/or hospitalization. However, home monitoring is not included in insurance schemes, whilst hospital in-patient costs are very high (around 40-110 euros daily, in a country with an average salary still below 100 euros per month).

Georgia had its first eHealth application in cardiology in 2000-2001, when coronarography files of patients with chronic IHD and acute coronary syndromes, like Miocardial Infarction were transferred from Chapidze Emergency Cardiology Center (ECC) for second opinion to Bad Krozingen Heart Center (Germany). The selected patients, who had to undergo not the easier stenting procedure (already available in Tbilisi) but the more serious coronary bypass surgery, were then sent to Germany until 2002. After that, the operation also became routine at CECC [9].

Another project, the first mHealth service in Georgia and funded by a local branch of US CRDF (Civilian Research and Development Foundation) and local NSF (National Science Foundation) funds, has tested the feasibility of tele-monitoring cardiac arrhythmia for patients outside the hospital [10]. The Vitaphone (Germany) ECG loop recorder and special software pre-recorded GSM phones have been used [6]. The local mobile operator provided his services for free for the arrhythmia monitoring cases. The ECG Loop Recorder automatically recorded arrhythmia events, which were transmitted via Bluetooth to the mobile phone and then by 3G or GPRS (partner mobile operator - MagtiCom Ltd.) to the Vita server in Germany and available to Georgian physician via e-mail/Internet. Cases of sinus brady- and tachyarrhythmia, sinus node weakness syndrome, atrial fibrillation, supraventricular tachycardia, supraventricular premature complexes (SPCs) and ventricular premature complexes (VPCs) have been correctly recognized by automatic recognition software and recorded. Patient observations by mobile telemedicine turned out 2.5-5 times less expensive than keeping them at cardiac wards. 85% of the patients have assessed the technology as very comfortable.

In a second stage, the above-mentioned project aims to be extended to Arterial Hypertension (2012) and later – to Chronic Heart Failure patients. In addition, starting from 2012, eHealth applications for cardiac patients will be provided to the Adjara Healthcare system in order to stay connected with some mountain villages that are out of reach in winter due to heavy snow. For these villages, tele-consultations are extremely important.

Relative challenges for business applications are related to high hardware cost and yet a needed paradigm shift in insurance schemes. In general, insurance companies have the policy to cover high inpatient costs, but are hesitating so far to support outpatient monitoring costs. The issue is discussed with the Parliamentary commission on health and social care and some changes are expected. Another issue is the fee for physicians who monitor the patient's ECG 24 hour. So far only four centers have enrolled in the program and due to relatively low number of patients, costs for services per participant, especially without insurance support, are high.

4.2 Actual Status: Belgium

Although the population of Belgium is rapidly aging and pressure on the care sector is still rising, this developed Western European country has not succeeded yet to implement any kind of generally accepted and applied model of a heart failure monitoring application. The main causes of death are also related, as practically all Western Countries, to cardiovascular problems. The big issue is that the number of cardiovascular patients grows exponentially starting from the age of 60 [5].

Now policy makers realize that tele-monitoring applications can be a key answer to the challenge of Belgium's aging, and more care demanding population. We noted that several projects, involving tele-monitoring patients, were done or are running at this moment [18]. Five projects are handling about monitoring heart failure patients; two of them are already finished. Both studies proved better quality of life for the patient and reduced cost for the health insurances [4].

After interviewing cardiologists who led the projects, we had to conclude that the demand for monitoring heart failure patients is huge: 15000 new patients a year in Belgium. The method proved to be efficient. There are several hardware manufacturers and distributors, and the care sector itself that is thrilled to deploy such an application. Then why didn't a general rollout happen yet? It seems that the answer can be found in the actual business model. Today, tele-monitoring applications aren't recognized by the insurance companies as an official manner to monitor a patient's health condition. So except when one joins a research project, all the hardware has to be bought by the patient or his/her family. Next to that, the professional care givers need time to analyze the results when an alarm is triggered or generated without being paid for it. In fact they lose income because in that time, other patients could have been examined. Another fundamental question that still needs to be answered handles about who needs to analyze and monitor the results. A GP has a more trust based relationship with his/her patients, but has normally only a few of chronic heart failure patients. So one can think that a specialized team of heart failure nurses, which can also build a sort of trust relationship with the patients, is more efficient then a GP, but then a GP loses contact with his patients, so an intermediate solution should be created.

So in order to implement a generally accepted heart failure tele-monitoring application one important final step should be made. More investigation is needed for the individual business models for the actors in order to achieve a social optimum. There has to be a way that the system is affordable and that all actors involved earn according to their involvement. The federal institution for healthcare insurance (RIZIV) is now working on the rollout of a so called care path. A care path is an individual approach for offering the right kind of healthcare to a patient. In this system there is an incentive foreseen per patient for all professional actors involved such as the GP, the cardiologist, hardware seller.

4.3 Actual Status: Cyprus

In Cyprus, there are various activities that utilize modern telecommunication technologies in an attempt to improve health-related services. These telemedicine applications range from home monitoring of patients to the enhancement of emergency response services.

In the home monitoring front, the focus is on children with cardiac arrhythmias which constitute one of the most difficult problems in cardiology both in terms of diagnosis and management. As said, the continuous monitoring of ECG vital signs and environmental conditions can significantly improve the chances of identifying possible arrhythmias. The outcome of this work is the prototype of a system [12] that is specifically designed for children and is able to carry out the acquisition and transmission of ECG signals. The fact that data transmission takes place in real-time allows for the deployment of an alarm-type scheme that assists in the identification of possible arrhythmias by notifying the on-call doctor and the child's relatives when abnormal events take place.

Turning the focus to emergency health care support, one activity aims in enhancing support to medical staff handling an emergency situation. The system developed [11] is a combined real-time and store and forward facility that consists of a base unit and telemedicine (mobile) units. While at the emergency site (e.g., ambulance, rural health centre, or ship) the mobile telemedicine unit communicates with a base unit at the hospital-expert's site. The system allows for the real time transmission of vital biosignals (3-12 lead ECG, SPO2, NIBP, IBP, and temperature) and still images of the patient. Its use was demonstrated and tested in Greece, Cyprus, Italy and Sweden [16]. The system is deployed in the district of Paphos in Cyprus: a rural health center, a small hospital and an ambulance are all connected to the Paphos General Hospital.

To provide further assistance in emergency situation handling, other efforts concentrate on enhancing the response times of the emergency service teams in cases of new incidents since in a number of cases any unnecessary delay may prove to be critical. In an attempt to start incorporating features that relate to Next Generation 112/911 services, smart call management mechanisms have been suggested [13] that allow the PSAP (Public Safety Answering Point) to apply call priority policies (e.g. for calls coming in from new numbers). Through such smart call handling algorithms and exploiting the features of today's smart phones for exact location identification one may reduce the response time to an incident. Moreover, analyzing past calling history in combination with location information can provide an estimate on the extent of an incident thus allowing for better resource planning and allocation.

The solutions described so far operate on the assumption that a modern telecommunications infrastructure is in place and constantly operational. Unfortunately, this is not the case worldwide and especially in some rural areas and large parts of developing countries. To allow for the transfer of telemedicine data under such conditions, a system [19] has been devised that operates according to the store-and-forward principle utilized in Delay Tolerant Network (DTN) deployments. The particular implementation concentrates on the transfer of low volume data that can be quickly read or easily keyed-in on a telephone keypad by a patient or caregiver and is not time critical e.g. scheduled blood pressure readings.

The potential outcomes of the work described in this section are highly promising and this has been demonstrated through small scale deployments. However, to date, a country-wide deployment of a telemedicine-related platform is still not present.

5 Comparison between Countries

Georgia, Belgium and Cyprus and most Western Countries would profit immensely by rolling out a tele-monitoring application for heart failure or cardiac arrhythmia patients. But emphasizes are different.

- In Belgium, already a backbone of care infrastructure for heart failure patients is foreseen, such as specialized heart failure hospital departments. The main reason of developing such a service would be to lower the cost for healthcare system and increase the quality of life.
- In Georgia, the health sector is still under revision. The primary goal is to reach people in rural areas and to provide stimuli to develop primary health care and outpatient tele-monitoring. Lowering costs is important, but to do that efficiently, medical management programs (such as HFMDP, Arterial Hypertension Control program) should be generalized and optimized first.
- Cyprus lies in between. A dedicated heart center is not a reality yet, but they're integrating one. On top of that, according to [26], they tend to possess a lot of expertise and awareness of the possibilities of eHealth and mHealth.

It may be concluded that, for these three countries, 'the last mile' is the major drawback to implement mHealth applications to monitor the heart condition. Although several projects proved to have successful results, Belgian, Cypriot and Georgian health insurers tend to have a very cautious view. They are willing to pay huge amounts of money to inbound patients in hospitals, but a much more economical solution such as tele-monitoring is not paid back. Fortunately, negotiations with small private insurers are taking place. Hopefully this is a trigger for other, more incumbent, health insurance companies. Once a good acceptance is created, other mHealth applications can be implemented as well.

6 Conclusions and Lessons Learned

The following conclusions can be drawn:

- The reference case showed that all countries (less/well developed) can gain by implementing a tele-monitoring service. Not only for lowering costs, but also for improving population's access to healthcare services.
- A sustainable business model that takes into account all the individual interest of the actors involved, professional and unprofessional (Cardiologist, general practitioner, heart failure nurses, health insurance companies, policy makers,…) is the key to a generally accepted tele-monitoring program [22].
- There exists a huge market potential for tele-monitoring services. In order to support this growth, actors like mobile operators, health insurance companies, hardware and application developers, etc. should further innovate and adapt their actual products. Next to these regulated services, non- regulated applications in the form of smart phone and internet applications will also become a part of our daily life.

In general can be said that the introduction of smart ICT and Future Internet services, such as video telephony, telemedicine, tele-monitoring, and personal alarm systems can contribute to a more sustainable and qualitative life for the whole population with emphasize on the elderly people living at home. This involves a lot of actors in a complex eco-system, all with their own expectations and outlook for potential benefit, either in business or social benefit.

That is the reason why it is so important to succeed in implementing the tele-monitoring service for heart failure patients. All necessary structures are already available, it's 'just' a matter of formulating a sustainable business model for all actors and the resulting Snowball effect will speed up the implementation of all following mHealth applications.

References

1. Dadoo, V.: White Paper on Mobile Health (2006), http://www.interactware.com/wp-content/uploads/2011/08/White-paper-Mobile-Healthcare-2006.pdf
2. Dang, S., Dimmick, S., Kelkar, G.: Evaluating the Evidence Base for the Use of Home Telehealth Remote Monitoring in Elderly with Heart Failure. Telemedicine and e-Health 15, 783–796 (2009)

3. Dansky, K.H., Palmer, L., Shea, D., Bowles, K.H.: Cost Analysis of Telehomecare. Telemedicine Journal and e-Health 7, 225–232 (2001)
4. Dendale, P.: Intensive telemonitoring program of patients with chronic heart failures (2010), http://www.jessazh.be/deelwebsites/Hartcentrum_Hasselt/Wetenschappelijke_info/Cardiologisch_congres_2010/Cardio_2010_Dendale_Telemonitoring_in_hartfalen
5. K. M. O. FOD Economie, Middenstand en Energie, Causes of Death, statistics for 2006, http://statbel.fgov.be/nl/binaries/NL%20-%20Tableau%201.3_T__tcm325-138394.pdf
6. Fransen, E.: The Vitaphone Tele-ECG System: Event Recording in Patients With Arrhytmias (2009), http://www.slideshare.net/guestf9a3fbf/vitaphone-tele-ecg-system-presentation
7. I. Global Industry Analysts, Global Telemedicine Market to Exceed $18 Billion by 2015 (2009)
8. Hebert, K., Quevedo, H.C., Gogichiashvili, I., Nozadze, N., Sagirashvili, E., Trahan, P., Kipshidze, N., Arcement, L.: Feasibility of a Heart Failure Disease Management Program in Eastern Europe: Tbilisi, Georgia. In: Circulation: Heart Failure, September 7 (2011)
9. Kirtava, Z., Aladashvili, A., Chapidze, G.: Still Image Teleconsulting for Interventional Cardiology in Georgia. Presented at the Abstracts of the 6th Int. Conference on Telemedicine, vol. 9, pp. 342–343 (2001)
10. Kirtava, Z.: e-Health and m-Health: Trends and visions for resource-limited environment in developing economies - case of Georgia. Presented at the ESF and COST High-Level Conference "Future Internet and Society – Complex Systems Perspective", Acquafredda di Maratea, Italy (2010)
11. Kyriacou, E., Pavlopoulos, S., Berler, A., Neophytou, M., Bourka, A., Georgoulas, A., Anagnostaki, A., Karayiannis, D., Schizas, C., Pattichis, C., Andreou, A., Koutsouris, D.: Multi-purpose HealthCare Telemedicine Systems with mobile communication link support. BioMedical Engineering OnLine 2 (2003)
12. Kyriacou, E., Pattichis, C., Hoplaros, D., Kounoudes, A., Milis, M., Jossif, A.: A System for Monitoring Children with Suspected Cardiac Arrhythmias - Technical Optimizations and Evaluation. In: Bamidis, P.D., Pallikarakis, N. (eds.) XII Mediterranean Conference on Medical and Biological Engineering and Computing, vol. 29, pp. 924–927. Springer, Heidelberg (2010)
13. Lambrinos, L., Djouvas, C.: Using location information for sophisticated emergency call management. In: Lin, J.C., Nikita, K.S. (eds.) MobiHealth 2010. LNICST, vol. 55, pp. 227–232. Springer, Heidelberg (2011)
14. Lee, I.: in personalinformatics.org, ed: HCII, Carnegie Mellon University (2011)
15. Mikalajunaite, E.: 500m people will be using healthcare mobile applications in 2015, October 19 (November 2010), http://www.research2guidance.com/500m-people-will-be-using-healthcaremobile-applications-in-2015/
16. Mougiakakou, S., Kyriacou, E., Perakis, K., Papadopoulos, H., Androulidakis, A., Konnis, G., Tranfaglia, R., Pecchia, L., Bracale, U., Pattichis, C., Koutsouris, D.: A feasibility study for the provision of electronic healthcare tools and services in areas of Greece, Cyprus and Italy. BioMedical Engineering OnLine 10, 49 (2011)
17. Quantified Self (2011), http://quantifiedself.com/guide/tag/gadget/reviews/4
18. RIZIV, Telemonitoring voor chronische patiënten : financiering van 6 projecten, October 18 (2011), http://www.riziv.be/care/nl/infos/telemonitoring/index.htm#2

19. Scholl, J., Lambrinos, L., Lindgren, A.: Rural Telemedicine Networks using Store-andforward VoIP and DTN. Presented at the Medical Informatics Europe (MIE 2009), Sarajevo, Serbia (2009)
20. Seto, E.: Cost Comparison Between Telemonitoring and Usual Care of Heart Failure: A Systematic Review. Telemedicine and e-Health 14, 679–686 (2008)
21. Valeri, L., Giesen, D., Jansen, P., Klokgieters, K.: Business Models for eHealth. Presented at the ICT for Health Unit, DG Information Society and Media (2010)
22. Van Ooteghem, J., Ackaert, A., Verbrugge, S., Colle, D., Pickavet, M., Demeester, P.: Economic viability of eCare solutions. In: eHealth 2010: Proceedings of the Third International ICST Conference on Electronic Healthcare for the 21st Century (2010)
23. Whited, J.D.: Economic Analysis of Telemedicine and the Teledermatology Paradigm. Telemedicine and e-Health 16, 223–228 (2010)
24. Winblad, I., Hämäläinen, P., Reponen, J.: What Is Found Positive in Healthcare Information and Communication Technology Implementation?- The Results of a Nationwide Survey in Finland. Telemedicine and e-Health 17, 118–123 (2011)
25. World Health Organization, World Health Report 2002: Reducing risks, promoting healthy life. WHO Press, Geneva, Switzerland (2002)
26. World Health Organization, Surveillance of chronic disease: risk factors: countrylevel data and comparable estimates (SuRF reports; 2). WHO Press, Geneva, Switzerland (2005)
27. World Health Organization: Global Observatory for eHealth series 1 (2010), http://www.who.int/goe/publications/ehealth_series_vol1/en/index.html
28. World Health Organization Regional Office for Europe, Assessment of NCD prevention and control in Primary Health Care, Georgia (2009)

Social Communication Behaviors of Virtual Leaders

Daphna Shwarts-Asher

"Sarnat" School of Management, The College for Academic Studies, Israel
daphna@mla.ac.il

Abstract. More and more organizations are adapting the solution of e-teams - teams that can span distances and times to take on challenges that most local and global organizations must address. This experimental study examined leadership in the context of traditional teams using face-to-face communication and virtual teams using computer-mediated communication. The research question is which leadership functions are necessary to promote virtual team performance. A model, suggesting that leader communication behaviors mediate the relationship between "virtuality" and "team's outputs" will be presented, and a methodology to examine this model will be illustrated. The findings show that face-to-face team's output is partially superior to a virtual team's output, and that social communication behaviors of face-to-face leaders are positive than social communication behaviors of virtual leaders. Virtual team is a common way of working, and will expand in the future. Thus, the importance of the theoretical and practical implementation of the virtual leadership will be discussed.

Keywords: Virtual teams, leadership, communication behaviors, team performance.

1 Introduction

The number of virtual teams is increasing in today's workplaces. In virtual teams, the members can have different cultural backgrounds; they often work in different countries and are professionals in their own fields [36]. A virtual team has been emerging as an appealing, effective means to help organizations achieving their goals, because of its distinctive capabilities of overcoming traditional organizational barriers (e.g., cost, location, time, space, a lack of talents and expertise in an organization, etc.) to facilitate collaboration among different functions and establish strategic partnerships/alliances outside their boundaries [13]. While there is a growing body of research on knowledge and information economy issues and the changing sociology of work, empirical work specifically on virtual team operation is embryonic [16]. The unique aspects of virtual teams generate major barriers to their effectiveness. Are there ways in which these may be either overcome or mitigated?

Virtual teams present a new challenge to the concept and practice of leadership. The traditional ideas of leadership in teams are built on a foundation of face-to-face contact. Such leadership has a significant relational component, including building

A.M. Hadjiantonis and B. Stiller (Eds.): Telecommunication Economics, LNCS 7216, pp. 67–77, 2012.

trust, handling conflict, and dealing with sensitive issues [47]. In light of the growing phenomenon of virtual teams, the traditional definition of "leadership" will be discussed, as part of the model that predicts the influence of the virtuality on leadership processes, social and tasks, that effect team output. Finally a methodology will be illustrated to examine the research model and a discussion of preliminary finding. The research will contribute a better understanding of virtual teams' leadership in hope of improving the teams work in the virtual world.

2 Literature Review

As the wired world brings everyone closer together, at the same time as they are separated by time and distance, leadership in virtual teams becomes even more important. Information technology makes it possible to build far-flung networks of organizational contributors, although unique leadership challenges accompany their formation and operation [10]. A review of "Leadership" and "Virtuality" as two main dimensions will be described in the next section.

2.1 Leadership

Leadership, particularly transformational and empowering leadership, improves team performance [38], yet it seems that leadership has an indirect impact on team performance. The literature implies that leadership can foster team performance through mediators. Schaubroeck [34] argued that transformational leadership influenced team performance through the mediating effect of team potency. The effect of transformational leadership on team potency was moderated by team power distance and team collectivism, such that higher power distance teams and more collectivistic teams exhibited stronger positive effects of transformational leadership on team potency. Kearney [21] had also suggests that transformational leadership can foster the utilization of the potential, but frequently untapped, benefits entailed by both demographic and informational/cognitive team diversity. Nemanich [27] argued that transformational leadership behaviors promote ambidexterity at the team level. They also found support for the association between transformational leadership and learning cultures. Shin [35] found that transformational leadership and educational specialization heterogeneity interacted to affect team creativity in such a way that when transformational leadership was high, teams with greater educational specialization heterogeneity exhibited greater team creativity.

Burke [6] indicated that specific leadership behaviors were generally related to team performance outcomes. Most notably, empowerment behaviors accounted for nearly 30% of the variance in team learning [6]. Srivastava [37] showed that empowering leadership was positively related to both knowledge sharing and team efficacy, which, in turn, were both positively related to performance.

Team-centric view of leadership raises many team leadership functions that help teams in the service of goal accomplishment [26]. It is reasonable that team dynamics is one of the major mediators that can improves team performance. Kanaga [19] claimed that one of the dimensions of effective teams is positive internal relationships. Hultman [18] argue that deep teams (teams that surface implicit, intangible thoughts and feelings) will outperform shallow teams (teams that makes inference about explicit, tangible behavior without clarifying its accuracy) in terms of bottom-line results, because they have access to a more complete range of relevant information. Sarin [33] presented an examination of the effect of team leader characteristics on an array of conflict resolution behavior, collaboration, and communication patterns of teams. Their findings suggested that participative management style and initiation of goal structure by the team leader exert the strongest influence on internal team dynamics. Both these leadership characteristics had a positive effect on functional conflict resolution, collaboration, and communication quality within the team while discouraging dysfunctional conflict resolution and formal communications.

Leader behaviors are significant predictors of performance [1], apparently because team leaders often provide incentives for cooperation [15]. A challenging question is how different incentive schemes and their actual choice by the leader contribute to the team's success? Despite the increased work on leadership in teams, there is a lack of integration concerning the relationship between leader behaviors and team performance outcomes [6]. Burke [6] results indicate that the use of task-focused behaviors is moderately related to perceived team effectiveness and team productivity. Person-focused behaviors were related to perceived team effectiveness, team productivity, and team learning. Thamhain [41] claimed that while effective management of the technical aspects of the project is critical to success, team leaders must also pay close attention to managing relations across the entire work process, including support functions, suppliers, sponsors, and partners.

2.2 Virtuality

The virtuality level of a team has become an integral part of a team's definition [24]. Many variables are affected by the virtuality level of a team. Face-to-face team member are more cohesive [17], have stronger social ties [43], are more dedicated to the task and to other team members [30], have a stronger team identity [5] and have more affection to other team members [44], than in virtual teams. Strong social ties in virtual teams can be achieved but will take longer time than in face-to-face teams [8].

Many researchers have attempted to find the reasons why virtuality has a negative influence on team output: frequency and distance [12], the fact that team members are not familiar with one another [14], the difficulty in sharing information, and insufficient and confusing discussions [42]. Another group of researchers compared communication technologies, assuming that technology limits information [39]. The comparisons concluded that face-to-face teams are more efficient than teams using video [2], and video communication is more efficient than audio [7], adding text into video or audio communication improves performance [3], and satisfaction [29]. Maruping [25] show that teams tend to use different sorts of communication technologies for different kinds of interpersonal interactions.

2.3 Virtual Leadership

The issue of leadership in virtual teams is an increasingly important one for many modern organizations [9]. Getting a group of people to work successfully as a team - communicating effectively, establishing trust, sharing the load, and completing tasks on time - is difficult even when the team members are all in the same location. When team members are spread out in various locations, it presents new obstacles for the team leader [23]. Virtual teams rely on computer-mediated communication, and team members have to find ways to deal with leadership using relatively lean media. Virtual team interaction occurs across the boundaries of geography, time, culture, organizational affiliation and a whole host of other factors. Many questions relating to virtual team leadership arise, including how well team members can express roles across distance and time, and what the role of facilitators is in virtual teams [47]. As such teams communicate mainly through communication technology this raises the challenge for the team leader of how to unify the team and get the members to identify themselves with the team [36].

While the behavioral and trait approaches are dominant in explaining effective leadership, contingency leadership theories must be considered explaining effective virtual leadership. Purvanova [31] results suggest that transformational leadership has a stronger effect in teams that use only computer-mediated communication (in compare to traditional teams), and that leaders who increase their transformational leadership behaviors in such teams achieve higher levels of team performance. Konradt [22] showed that middle managers compared to line managers perceived people oriented leadership functions (i.e., mentor and facilitator roles) as important whereas line managers compared to middle managers perceived stability leadership functions (i.e., monitor and coordinator roles) as important. Nicholson [28] found that face-to-face and cross-cultural virtual team-members value different ingredients of leadership in different phases of the project.

Zaccaro [46] examined the similarities and differences between virtual teams and face-to-face teams. They suggest that affective processes include the expression of emotion by e-team members, as well as the management of these expressions. Eom [13] argued that trust is a key proxy for a virtual team's success, since trust enhances the performance of a virtual team. Carte [9] results suggest that high performing virtual teams displayed significantly more leadership behaviors over time compared to their low performing counterparts. Specifically, these teams displayed significantly more concentrated leadership behavior focused on performance (i.e. "Producer" behavior) and shared leadership behavior focused on keeping track of group work (i.e. "Monitor" behavior) than the lower performing teams.

There are aspects of virtual team leadership that may help overcome some of the potential process losses associated with virtual teamwork [11]. The most salient challenges for E-leaders of virtual teams are the difficulty of keeping tight and loose controls on intermediate progress toward goals [10]. Kayworth [20] suggest that effective team leaders demonstrate the capability to deal with paradox and contradiction by performing multiple leadership roles simultaneously (behavioral complexity). Specifically, it is discovered that highly effective virtual team leaders act in a mentoring role and exhibit a high degree of understanding (empathy) toward other team members. Sivunen [36] study focuses on four virtual team leaders and their attempts to strengthen the team members' identification with the team through computer-mediated communication. The results show four different tactics employed in enhancing identification with the team: catering for the individual, giving positive feedback, bringing out common goals and workings and talking up the team activities and face-to-face meetings.

3 Research Model

The research model is depicted in Figure 1. According to the model the **virtuality level** is an affecting variable, while the measurable (dependent) variables are the outputs of the team work: **efficiency, effectiveness** and **satisfaction**. The **leader communication behaviors** are variable which mediate the relationship between Virtuality and Team's outputs.

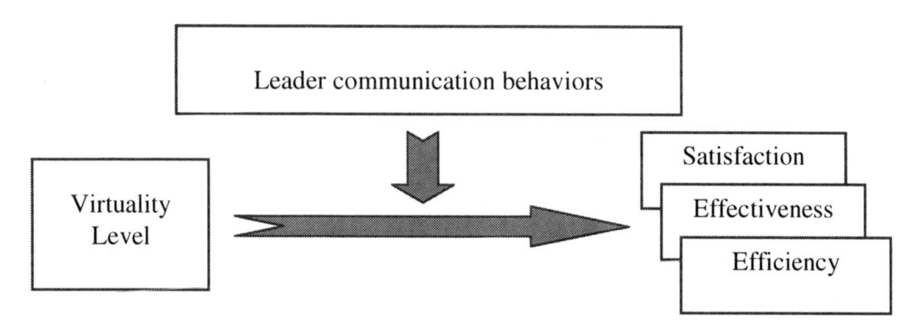

Fig. 1. Research Model

4 Research Hypotheses

- Hypothesis 1 – for an intellective task, social communication behaviors of face-to-face leaders are positive than social communication behaviors of virtual leaders.
- Hypothesis 2 – for an intellective task, social communication behaviors of virtual leaders are negative than social communication behaviors of face-to-face leaders.
- Hypothesis 3 – for an intellective task, task communication behaviors of face-to-face leaders include more answers than task communication behaviors of virtual leaders.
- Hypothesis 4 – for an intellective task, task communication behaviors of virtual leaders include more questions than task communication behaviors of face-to-face leaders.
- Hypothesis 5 – for an intellective task, face-to-face team's output is superior to a virtual team's output.

5 Methodology

An experiment was designed, in which a team task was delivered to 75 undergraduate students in an academic college. The subjects, who were grouped into teams of three members, had to share information in order to complete the task. Each team was given a task that takes approximately 30 minutes to complete. The research design is a Between Subjects Factorial Design: the factor is the type of communication: virtual vs. face-to-face. The research design includes a total of two experimental conditions. The virtual condition was implemented on 13 teams, while the Face-to-face condition was implemented on 12 teams, as described in Table 1. Thus, the experiment included 75 subjects (2 conditions * 12-13 teams * 3 subjects).

Table 1. Experimental Conditions

	Virtuality Level	N	Remarks
1	0	12	Face-to-face team
2	1	13	virtual team

5.1 Procedure

Subjects were invited in groups of three to meetings that were conducted using MSN-Messenger (virtual) or face-to-face (non virtual) communication. At the beginning of the meetings, the team members were asked to nominate a chairperson. The process of the experiment includes an intellective task. Each team member received a discrete and different piece of information, and only the aggregation of all the information revealed the whole "picture" and led to the correct solution.

5.2 Operationalization of Dependent Variables and Mediators

- ***Efficiency*** – the time required to complete the task.
- ***Effectiveness*** – the team's solution compared to the correct solution.
- ***Satisfaction*** – team members' reaction to the task will be measured by their understanding of communication, and satisfaction of medium, results and process.
- ***Leader communication behaviors*** – A textual (or audio) recording was saved for each virtual (or face-to-face) meeting. Task and social communication behaviors of leaders were measured by content analysis: The analysis, for each team leader (and actually for each team member) at any meeting, included the number of social positive phrases, social negative phrases, task question phrases and task answer phrases, accordingly to Bales [4] model. In order to use reliable measures, the phrase counting was done separately by two independent judges. The two judgment analysis was compared one to the other, and in a case of different decision (concerning the phrase category), a new agreed decision was taken.

Four measures were calculated out of the above phrases counting:

- Positive (Social) Leadership Level – Positive phrases percentage among all phrases of the leader during the meeting.
- Negative (Social) Leadership Level – Negative phrases percentage among all phrases of the leader during the meeting.
- Questions (Task) Leadership Level – Question phrases percentage among all phrases of the leader during the meeting.
- Answers (Task) Leadership Level – Answers phrases percentage among all phrases of the leader during the meeting.

Table 2 presents a summary of the Means and SD's of all the mediating variables above the experiment condition (Means and SD's by the independent variables are described in the following section).

Table 2. Means and SD's of the Mediator Variables (Overall N=25)

Mediator Variables	*M*	*SD*
Positive (Social) Leadership Level	18%	6%
Negative (Social) Leadership Level	8%	8%
Answers (Task) Leadership Level	52%	8%
Questions (Task) Leadership Level	21%	4%

Table 3 presents a summary of the Means and SD's of all the output variables above the experiment condition (Means and SD's by the independent variables are described in the following section).

Table 3. Means and SD's of the Output Variables (Overall N=25)

Output Variables	*M*	*SD*
Success (Effectiveness)	83%	28%
time (Efficiency)	33.32	13.63
Satisfaction	3.80	0.38

6 Findings

Twenty five experiments where preformed (out of the 50 planned) among undergraduate students in an academic college. Table 4 presents a summary of the Means and SD's by the independent variables. Table 5 presents a summary of the Means and SD's of all the output variables by the independent variables. A statistical analysis was performed. The T-Tests conducted compared communication behaviors of face-to-face leaders to communication behaviors of virtual leaders, and face-to-face team's output is to virtual team's output. The analysis indicates, for each hypothesis respectively:

Table 4. Means and SD's of the Mediator Variables by the independent variables

Mediator Variables	virtual leaders (N=13)		f-t-f leaders (N=12)	
	M	*SD*	*M*	*SD*
Positive (Social) Leadership Level	15.06%	4.69%	22.23%	4.32%
Negative (Social) Leadership Level	9.94%	11.71%	6.50%	2.29%
Answers (Task) Leadership Level	21.80%	5.47%	19.84%	3.19%
Questions (Task) Leadership Level	21.80%	5.47%	19.84%	3.19%

Table 5. Means and SD's of the Output Variables by the independent variables

Output Variables	virtual leaders (N=13)		face-to-face leaders (N=12)	
	M	*SD*	*M*	*SD*
Success (Effectiveness)	86.54%	26.25%	78.47%	29.40%
time (Efficiency)	41.85sec	12.55sec	24.08sec	8.46sec
Satisfaction	3.66	0.45	3.96	0.26

- H1 – social communication behaviors of face-to-face leaders are positive than social communication behaviors of virtual leaders ($t = 3.96$; $p < 0.05$).
- H2 – social communication behaviors of virtual leaders are not negative than social communication behaviors of face-to-face leaders.
- H3 – task communication behaviors of face-to-face leaders does not include more answers than task communication behaviors of virtual leaders.
- H4 – task communication behaviors of virtual leaders does not include more questions than task communication behaviors of face-to-face leaders.
- H5 – face-to-face team's output is partially superior to a virtual team's output: face-to-face teams are not successful than virtual teams in completing the task, yet for virtual teams it takes longer time in carrying out the task ($t = -4.11$; $p < 0.05$) and the virtual teams members are lees satisfied ($t = 2.04$; $p < 0.05$).

7 Discussion

Collaboration in distributed settings has become a reality in organizational life, while information flows freely across organizational, geographic, and cultural borders. More and more organizations are adapting the solution of e-teams - teams that can span distances and times to take on challenges that most local and global organizations must address [46]. This experimental study examined leadership in the context of traditional teams using face-to-face communication and virtual teams using computer-mediated communication, in order to check what leadership functions are necessary to promote virtual team success and performance, in light of Horwitz [16] claims about the importance of leadership communication to virtual team performance.

The results shows that face-to-face team's output is partially superior to a virtual team's output: while face-to-face and virtual teams are equally successful in completing the task, virtual teams takes longer time in carrying out the task and their members are lees satisfied.

Three out of four hypotheses concerning the team leader communication behaviors were refuted: social communication behaviors of virtual leaders are not negative than social communication behaviors of face-to-face leaders; task communication behaviors of face-to-face leaders does not include more answers than task communication behaviors of virtual leaders and; task communication behaviors of virtual leaders does not include more questions than task communication behaviors of face-to-face leaders.

Yet, it is possible that the significant difference between social communication behaviors of face-to-face leaders and virtual leaders can act as an explanation for the increased face-to-face team's output in compare to the virtual team's output. The results indicate that social communication behaviors of face-to-face leaders are positive than social communication behaviors of virtual leaders. Existing theory and research reveals that constructive management behaviors are important to teams' success. Wolff [45] contributes to existing theory by proposing that empathy precedes and enables those cognitive processes and skills by providing an accurate understanding of team and member emotions and needs. Rego [32] suggest that emotionally intelligent leaders behave in ways that stimulate the creativity of their teams. Tansley [40] showed that trust is a necessary pre-condition for the development and exploitation of social capital, a significant influence on project success.

Though the findings are, in general, consistent with the existing literature, it strengthens the importance of positive social communication behavior as a specific leadership communication behavior, rather than any other type of behavior. It also implies that leadership positive social communication behavior can explain the difference between face-to-face team's outputs in compare to the virtual team's output.

References

1. Ammeter, A.P., Dukerich, J.M.: Leadership, team building, and team member characteristics in high performance project teams. Engineering Management Journal 14(4), 3–11 (2002)
2. Andres, H.P.: A Comparison of Face-to-Face and Virtual Software Development Teams. Team Performance Management: An International Journal 8(1/2), 39–48 (2002)
3. Baker, G.: The Effects of Synchronous Collaborative Technologies on Decision Making: A Study of Virtual Teams. Information Resources Management Journal 15(4), 79–93 (2002)
4. Bales, R.F.: A set of factors for the analysis of small group interaction. American Sociological Review 15, 257–263 (1950)
5. Bouas, K.S., Arrow, H.: The Development of Group Identity in Computer and Face-to-face Groups with Membership Change. CSCW 4, 153–178 (1996)
6. Burke, C.S., Stagl, K.C., Klein, C., et al.: What types of leadership behaviors are functional in teams? A meta-analysis. Leadership Quarterly 17(3), 288–307 (2006)
7. Burke, K., Aytes, K., Chidambaram, L.: Media Effects on the Development of Cohesion and Process Satisfaction in Computer-Supported Workgroups. Information Technology & People 14(2), 122–141 (2001)

8. Burke, K., Chidambaram, L.: Do Mediated Contexts Differ in Information Richness? A Comparison of Collocated and Dispersed Meetings. In: 29th Annual Hawaii International Conference on System Sciences, Hawaii, USA, pp. 92–101 (1996)
9. Carte, T.A., Chidambaram, L., Becker, A.: Emergent Leadership in Self-Managed Virtual Teams; A Longitudinal Study of Concentrated and Shared Leadership Behaviors. Group Decision and Negotiation 15(4), 323–343 (2006)
10. Cascio, W.F., Shurygailo, S.: E-leadership and virtual teams. Organizational Dynamics 31(4), 362–376 (2003)
11. Cordery, J.L., Soo, C.: Overcoming impediments to virtual team effectiveness. Human Factors and Ergonomics in Manufacturing 18(5), 487–500 (2008)
12. Cramton, C.D., Webber, S.S.: Modeling the Impact of Geographic Dispersion on Work Teams. In: Working Paper, George Mason University, Washington, DC, USA (1999).
13. Eom, M.: Cross-Cultural Virtual Team and Its Key Antecedents to Success. The Journal of Applied Business and Economics 10(1), 1–15 (2009)
14. Gruenfeld, D.H., Mannix, E.A., Williams, K.Y., Neale, M.A.: Group composition and decision making: How member familiarity and information distribution affect process and performance. Organizational Behavior and Human Decision Processes 67(1), 1–16 (1996)
15. Gürerk, Ö., Irlenbusch, B., Rockenbach, B.: Motivating teammates: The leader's choice between positive and negative incentives. Journal of Economic Psychology 30(4), 591–607 (2009)
16. Horwitz, F.M., Bravington, D., Silvis, U.: The promise of virtual teams: identifying key factors in effectiveness and failure. Journal of European Industrial Training 30(6), 472–494 (2006)
17. Huang, W.W., Wei, K.K., Watson, R.T., Tan, B.C.Y.: Supporting Virtual Team-Building with a GSS: An Empirical Investigation. Decision Support Systems 34(4), 359–367 (2003)
18. Hultman, K., Hultman, J.: Deep Teams: Leveraging the Implicit Organization. Organization Development Journal 26(3), 11–23 (2008)
19. Kanaga, K., Browning, H.: Keeping watch: How to monitor and maintain a team. Leadership in Action 23(2), 3–8 (2003)
20. Kayworth, T.R., Leidner, D.E.: Leadership effectiveness in global virtual teams. Journal of Management Information Systems 18(3), 7–41 (2001/2002)
21. Kearney, E., Gebert, D.: Managing diversity and enhancing team outcomes: The promise of transformational leadership. Journal of Applied Psychology 94(1), 77–89 (2009)
22. Konradt, U., Hoch, J.E.: A Work Roles and Leadership Functions of Managers in Virtual Teams. International Journal of E-Collaboration 3(2), 16–25 (2007)
23. Kossler, M.E., Prestridge, S.: Going the distance: The challenges of leading a dispersed team. Leadership in Action 23(5), 3–6 (2003)
24. Martins, L.L., Gilson, L.L., Maynard, M.T.: Virtual teams: What do we know and where do we go from here? Journal of Management 30(6), 805–835 (2004)
25. Maruping, L.M., Agarwal, R.: Managing Team Interpersonal Processes through Technology: A Task-Technology Fit Perspective. Journal of Applied Psychology 89(6), 975–990 (2004)
26. Morgeson, F.P., DeRue, D.S., Karam, E.P.: Leadership in Teams: A Functional Approach to Understanding Leadership Structures and Processes. Journal of Management 36(1), 5–39 (2010)
27. Nemanich, L.A., Vera, D.: Transformational leadership and ambidexterity in the context of an acquisition. Leadership Quarterly 20(1), 19–33 (2009)
28. Nicholson, D.B., Sarker, S., Sarker, S., Valacich, J.S.: Determinants of effective leadership in information systems development teams: An exploratory study of face-to-face and virtual context. Journal of Information Technology Theory and Application 8(4), 39–56 (2007)

29. Olson, J., Olson, G., Meader, D.: Face-to-face Group Work Compared to Remote Group Work With and Without Video. In: Finn, K., Sellen, A., Wilbur, S. (eds.) Video-mediated Communication, pp. 157–172. Lawrence Erlbaum Associates, Mahwah (1997)
30. Olson, J., Teasley, S.: Groupware in the Wild: Lessons Learned from a Year of Virtual Collocation. In: Proceedings of the ACM Conference, Denver, CO, USA, pp. 419–427 (1996)
31. Purvanova, R.K., Bono, J.E.: Transformational leadership in context: Face-to-face and virtual teams. Leadership Quarterly 20(3), 343–357 (2009)
32. Rego, A., Sousa, F., Pina e Cunha, M., Correia, A., Saur-Amaral, I.: Leader Self-Reported Emotional Intelligence and Perceived Employee Creativity: An Exploratory Study. Creativity and Innovation Management 16(3), 250–264 (2007)
33. Sarin, S., O'Connor, G.C.: First among Equals: The Effect of Team Leader Characteristics on the Internal Dynamics of Cross-Functional Product Development Teams. The Journal of Product Innovation Management 26(2), 188–205 (2009)
34. Schaubroeck, J., Lam, S.S.K., Cha, S.E.: Embracing transformational leadership: Team values and the impact of leader behavior on team performance. Journal of Applied Psychology 92(4), 1020–1030 (2007)
35. Shin, S.J., Zhou, J.: When is educational specialization heterogeneity related to creativity in research and development teams? Transformational leadership as a moderator. Journal of Applied Psychology 92(6), 1709–1721 (2007)
36. Sivunen, A.: Strengthening Identification with the Team in Virtual Teams: The Leaders' Perspective. Group Decision and Negotiation 15(4), 345–366 (2006)
37. Srivastava, A., Bartol, K.M., Locke, E.A.: Empowering leadership in management teams: effects on knowledge sharing, efficacy and performance. Academy of Management Journal 49(6), 1239–1251 (2006)
38. Stewart, G.L.: A Meta-Analytic Review of Relationships between Team Design Features and Team Performance. Journal of Management 32(1), 29–55 (2006)
39. Straus, S.G., McGrath, J.E.: Does the medium matter? The interaction of task type and technology on group performance and member reactions. Journal of Applied Psychology 79(1), 87–98 (1994)
40. Tansley, C., Newell, S.: Project social capital, leadership and trust; A study of human resource information systems development. Journal of Managerial Psychology 22(4), 350–368 (2007)
41. Thamhain, H.J.: Team leadership effectiveness in technology-based project environment. Project Management Journal 35(4), 35–47 (2004)
42. Thompson, L.F., Coovert, M.D.: Teamwork Online: The Effects of Computer Conferencing on Perceived Confusion, Satisfaction, and Post-discussion Accuracy. Group Dynamics: Theory, Research, and Practice 7(2), 135–151 (2003)
43. Warkentin, M., Sayeed, L., Hightower, R.: Virtual Teams versus Face-to-Face Teams: An Exploratory Study of a Web-Based Conference System. Decision Sciences 28(4), 975–996 (1997)
44. Weisband, S., Atwater, L.: Evaluating Self and Others in Electronic and Face-to-Face Groups. Journal of Applied Psychology 84(4), 632–639 (1999)
45. Wolff, S.B., Pescosolido, A.T., Druskat, V.U.: Emotional intelligence as the basis of leadership emergence in self-managing teams. Leadership Qtly. 13(5), 505–522 (2002)
46. Zaccaro, S.J., Bader, P.: E-leadership and the challenges of leading e-teams: Minimizing the bad and maximizing the good. Organizational Dynamics 31(4), 377–387 (2003)
47. Zigurs, I.: Leadership in virtual teams: Oxymoron or opportunity? Organizational Dynamics 31(4), 339–351 (2003)

Policy Implications of Digital Television

Bartolomeo Sapio

Fondazione Ugo Bordoni, Viale del Policlinico 147, 00161 Roma, Italy
bsapio@fub.it

Abstract. This section is an exploratory study about factors affecting the usage of T-government services for payments through Digital Terrestrial Television (DTT) and how policy measures can influence adoption. The widely known predicting model of ICT user acceptance UTAUT (Unified Theory of acceptance and Use of Technology) is applied to recognize those factors affecting usage, exploiting a dataset coming from an Italian T-government project. T-government stands for a wide set of services addressed to citizens (e.g. about health, education, tourism, payment of bills), delivered by a Public Administration or a private provider, and accessible through Digital Television. One of the opportunities given by T-government is to promote the use of ICT-based public services by large groups of people (e.g., the elderly), who haven't got any Internet access or the required skills to use it.

Keywords: Digital Terrestrial Television, policy measures, T-government, Unified Theory of acceptance, Use of Technology.

1 European Policies toward Digital Terrestrial Television

The introduction of Digital Terrestrial Television (DTT) is derived from the process of implementing the recommendations of the European Union: while they preserve competition between terrestrial, satellite and cable platforms within the EU market, the recommendations provide forms of public intervention such as: funding for pilot and research projects, subsidies for the purchase of decoders for any platform in order to prevent the exclusion of low-income families from access to TV reception, subsidies to companies (to develop innovative digital services) and broadcasters (to compensate for the additional transmission costs due to the parallel broadcasting of analogue and digital signals, the so-called "simulcast phase"). In this context, the major European countries have adopted policies that have led to high rates of digitization in families. Support policies have aimed to provide continuity of service in each country, especially considering that most of the population (particularly the weakest layers) uses television as the main source of information.

Table 1 outlines various policies carried out in four major European TV markets: Italy, France, United Kingdom and Spain.

A.M. Hadjiantonis and B. Stiller (Eds.): Telecommunication Economics, LNCS 7216, pp. 78–88, 2012.

Table 1. European policies for Digital Television *(Source: Digital UK, Ofcom, Impulsa TDT, CSA, Agcom, DGTVi, Booz & Company)*

	Italy	**France**	**United Kingdom**	**Spain**
Subsidies to users	Regional subsidies to purchase DTT decoders in switch-off regions	100 million Euros (in 3 years) for families including elderly or disabled people 85 million Euros (in 3 years) for low income families	603 million pounds for the Digital Switch-over Help Scheme (subsidies to target families)	110 million Euros (in 2008 and 2009) for regional communication and support to vulnerable groups
Subsidies to broadcasters	33 million Euros (2007) to RAI for DTT	Reserved channels for traditional free-to-air analogical broadcasters, in exchange for the transition of broadcasting networks to digital	Renewal of licenses in exchange for the transition of broadcasting networks to digital	Contributions to increase TDT coverage (8.7 million Euros in 2008)
Industrial choices	Obligation of DTT tuner in iDTV DGTVi quality certification	Obligation of DTT tuner in iDTV Push towards HD TNT quality certification	Freeview quality certification	Push towards MHP decoders

2 T-Government Pilot Projects

T-government stands for a wide set of services addressed to citizens (e.g., about health, education, tourism, payment of bills), delivered by a Public Administration or a private provider, and accessible by Digital Television. One of the opportunities given by T-government is to promote the use of ICT-based public services by large groups of people (e.g., the elderly), who haven't got any Internet access or the required skills to use it. For those people the past experience with TV and remote control may be a key qualification to become effective users of the above mentioned services. In this way the inclusion in the benefits of information society of culturally disadvantaged people can be encouraged (e-inclusion).

On these topics field investigations have been developed in Italy in the framework of the T-government projects promoted by Fondazione Ugo Bordoni (FUB). One of the objectives of the six T-government projects co-funded by FUB was to experiment high interactivity T-government services, realized by DTT involving real users.

The main aim of this study was to investigate the use behavior during the diffusion of DTT services. A multidisciplinary approach, in particular the framework provided by the human factors discipline, was adopted in order to focus the attention on the variables affecting both usage-usability and socio-economic aspects. In such a perspective a predicting model of ICT user acceptance, i.e. the UTAUT model [6] [7], was applied to recognize those factors directly affecting the usage. As a second step, a micro simulation model was implemented to investigate the diffusion patterns of both Digital TV and T-government services. The adopted innovative methodology progressed through the following main steps by:

1. Collecting general information about governmental policies, current state of DTT penetration in Italy as well as other T-government field investigations;
2. Structuring the data gathered in the Italian pilot studies to feed the micro simulation model;
3. Identifying the most relevant factors affecting services usage;
4. Building up the micro simulation model; and
5. Generating scenarios about citizens' adoption and use behavior of DTT services.

Given the existing general policy constraints, different strategic scenarios of citizens' adoption and use of DTV services offered by a public provider were envisaged to be dependant upon the governmental decisions, i.e. tax reduction, public communication campaign, and analogue system switch-off date.

The experimented services belong to different application areas: demographics, utilities and fines, education, T-health, T-learning, employment, T-commerce, T-banking. In order to investigate usage, usability and socio-economical aspects of T-government services, an explorative field investigation was developed in each of the six projects.

The field investigations were realised using a common framework referring to human factors discipline [4]. The following main usability aspects related to interactive services were identified: perceived usefulness, perceived ease of use and attractiveness, training and user support (human support, user manual, support provided by DTT, call centre), user perception of technical disturbances and troubles (due to television signal, set top box, return channel), security and privacy perception (confidentiality of personal data, security of payments), impact of the equipment in the house, users' satisfaction about the service including the comparison of different channels to perform the same task (e.g., DTT versus Internet, DTT versus traditional office desk).

The adopted indicators for service utilisation level are [1]: frequency of use, time duration of the session, and kind of use (shallow or intensive). Socio-economical aspects include user profile (including income and social network information), TV and Internet usage (including other entertainment technologies), and scenarios (including interest levels, willingness to pay for equipment and services, decision factors).

Services developed in three of the six projects are still "on air" after the end of the project. These services have the common feature of being provided in the context of a public administration (municipality, regional administration, public schools).

In the following sections three pilot studies are introduced, analyzing the effect of policy measures on the diffusion. Methodological issues are discussed in detail in the referenced papers.

2.1 Pilot Study "Parma Municipality Services"

The project "Parma Municipality Services" [5] has developed both informative and interactive services. The former provide information about: Parma Municipality organisation; services offered to the citizens by the Municipality; cultural initiatives in progress in the city.

The interactive services allow the user (after the authentication through the Electronic Identity Card):

- to pay fines using a credit card;
- to visualise the state of a demographic dossier (e.g., for changing place of abode);
- to visualise the state of a request for education services provided by the municipality (e.g., nursery school).

A sample of 200 people was selected from a group of 4.000 citizens of Parma Municipality. It was decided to choose the sample age ranging from 20 to 45 years, people in general skilled in the use of new information technology. The 200 citizens were randomly selected in this age range. Finally, 181 citizens (88 males and 93 females) took part in the experiment for the all planned period of two months and a half. Data were collected mainly using a paper-based questionnaire filled in by the users. The questionnaire was both delivered and collected by Municipality of Parma personnel.

Scenarios

Alternative reference scenarios are presented hereafter in Figure 1 and Figure 2. They are not intended to be predictive, but only to investigate the model mechanisms.

Scenario 1: Switch off in 2013, No subsidy, No communication. In the first year, all adopters do not use services. The year after, all of them use only information services. Then, throughout four years, the adopters are divided between those who use interactive and information services and those (the major part) who exploit only information services, and their total number remains constant and equal to the initial value. After the switch off, when all users become adopters, there is no increase of real users of services for one year. After that, a great rise of the users of information services can be observed and, a year later, a more limited increase of the users exploiting both information and interactive services occurs too, with a corresponding decrease of information service users.

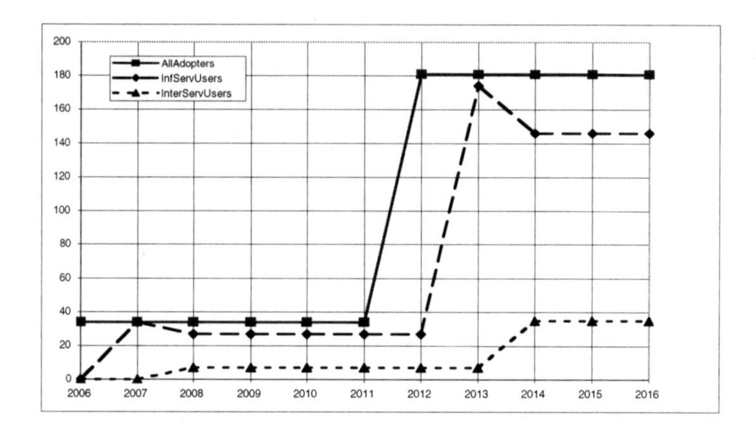

Fig. 1. Results from different policy scenarios in scenario 1

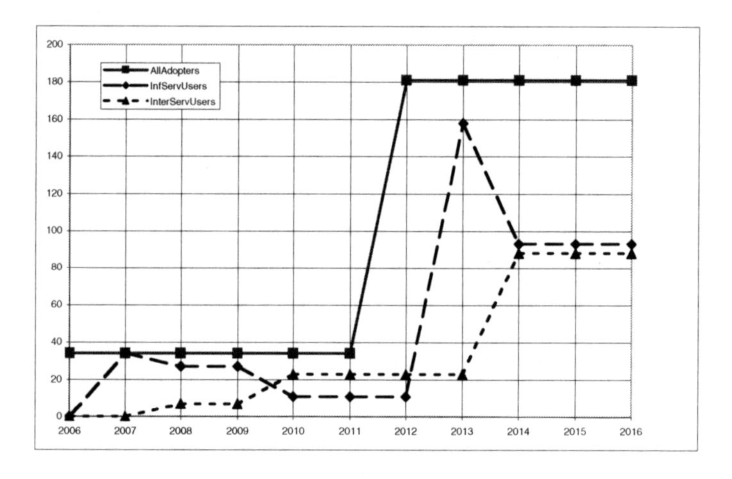

Fig. 2. Results from different policy scenarios in scenario 2

Scenario 2: Switch off in 2013, Communication (level 2) 2010-2014, No subsidy. The most important variation introduced by an intense communication lasting from 2010 to 2014 (in absence of subsidization) is in the sharing of users between information service users and users of both information and interactive services, thus, in the shaping of relevant lines. Two inversions of dominance can be observed before the two classes of users reach constant and very close final numbers of elements.

2.2 Pilot Study "Employment Services"

The project "Employment Services" [2] developed a DTV portal to provide support - especially to young people - in finding a job. The DTV portal was organized in four sections:

1. Search for a job
2. Search for information useful to find an employment

3. T-commerce. It was possible to buy books and to pay using a credit card
4. T-learning. Broadcast lessons were delivered

A panel of about 400 users was involved in the field study. The panel was distributed in the geographical areas of Pisa, Lucca and Livorno in Tuscany region. The users were selected with the criteria to be potentially interested in finding a job. As a consequence, they were belonging to the range of age in which is probable to search for an employment. The users were mainly recruited among the users participating in a previous study about DTV utilization, so that most part of them was equipped in advance with the set-top box.

Scenarios

Alternative reference scenarios are presented hereafter. In the scenario generation process, three different policies and combinations of them have been tested:

- ***Policy EE*:** design of DTV applications with increased usability to reduce the user's effort expectancy
- ***Policy PE*:** communication campaign to increase the user's performance expectancy from DTV services
- ***Policy FC*:** user support (through call centers and similar) to create facilitating conditions.

Figure 3 shows the adopters' curve as generated by the micro simulation model. It is not influenced by the different policies and only depends on Bass model parameters ($p = 0.014$, $q = 0.726$) [7].

The first DTT users in Italy came in 2004, the year of take-off. The adoption rate is slow during the first years and becomes progressively quicker when the switch-off date gets closer. As an intrinsic characteristic of the model, all users become adopters from 2013, when analog broadcast is no longer in operation. The dashed line represents the adoption as it would happen according to the "pure" Bass model, the full line includes also the effect of switch-off.

Scenario 1: No policies. The use behavior (mean number of monthly interactions) of T-government services without any governmental policies is shown in Figure 4. The use behavior has been obtained by the item of the questionnaire "frequency of use of the portal sections in the last month". The dashed lines respectively represent the standard deviation. In this scenario the growth is quite regular, slower during the first years and somehow quicker from 2009, without any singularities. Variations are not very significant and the mean number of monthly interactions remains relatively low during the entire time horizon.

Scenario 2: Combination of policies EE, PE and FC 2009-2020. Figure 5 shows a scenario where a combination of three different policies (EE, PE and FC) is implemented from the year 2009 to the end of the simulation period. This scenario conveys a significant increase of the monthly frequency of use during the implementation of the policies, reaching a nearly triple regime value. The standard deviation is quite significant in the transition period, suggesting that some users are reluctant to use T-government services while others are ready to exploit the new possibilities opened by digital television.

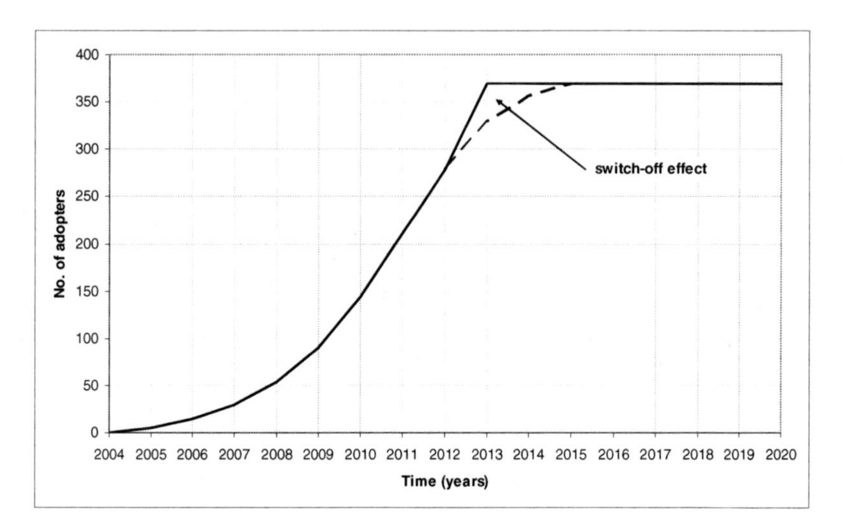

Fig. 3. Adopters' curve

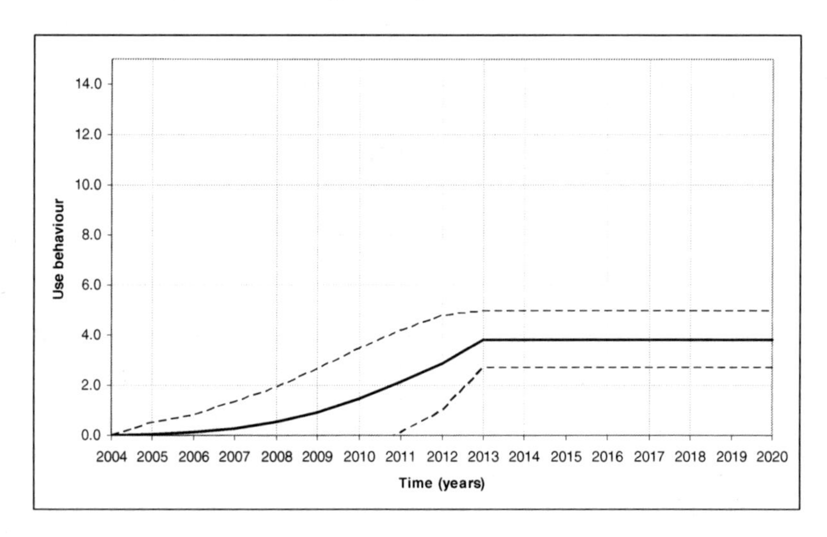

Fig. 4. Scenario 1 – No policies

2.3 Pilot Project "T-Bollettino"

T-Bollettino [3] is an interactive service providing the user with

- remote interactivity using the return channel
- user identification and authentication using a smart card
- on line payment functionality

The return channel is connected to the DTV service centre of the project, and the broadcasting system broadcasts the application.

The service allows bills' payment through DTV for gas, phone, local taxes, fines, road taxes. The payment can be realized using a credit card, a prepaid card (Postepay) or by charging the account "Bancoposta Online".

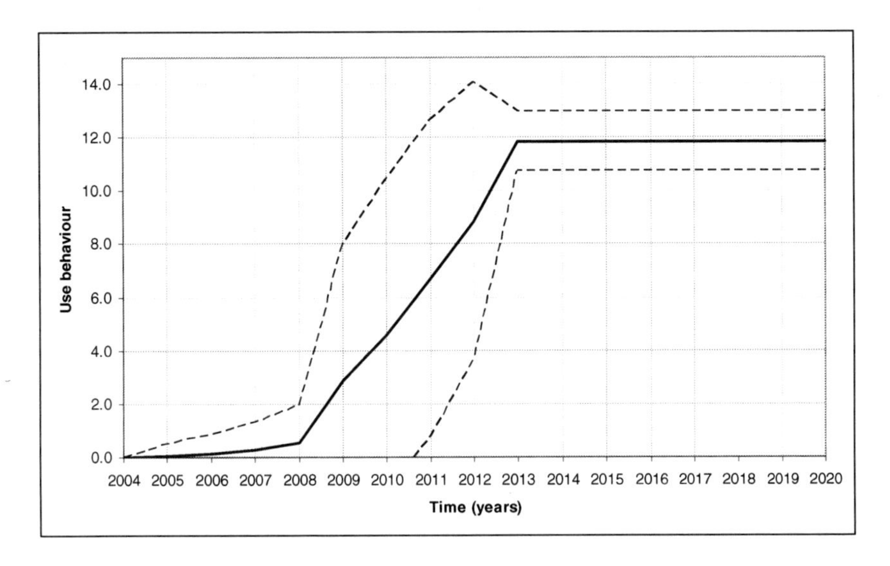

Fig. 5. Scenario 2 – Combination of policies EE, PE and FC

A panel of about 300 users was involved in the field study. The panel was uniformly distributed in the North, Centre (City of Roma) and South of Italy. The users were selected with the criteria to be representative of Italian families in terms of number of components of the family, age of head of the family, education of head of the family, job of head of the family. Only people usually paying bills at the post office and owners of a credit card, a prepaid card (Postepay) or the current account "Bancoposta" were included in the users' panel.

Scenarios

The impact of three different policies and combinations of them has been tested:

- ***Policy PE*:** communication campaign to increase the user's performance expectancy from DTV services
- ***Policy FC*:** user support (through call centers and similar) to create facilitating conditions.
- ***Policy PS*:** communication campaign or security interventions on applications to increase the users' perceived security

The simulation model allows to test the impact of two different intensity levels for each policy (level 1 = moderate, level 2 = strong). Policies based on effort expectancy (e.g., design of DTV applications with increased usability) have not been considered, since the effort predictor had a non-significant impact.

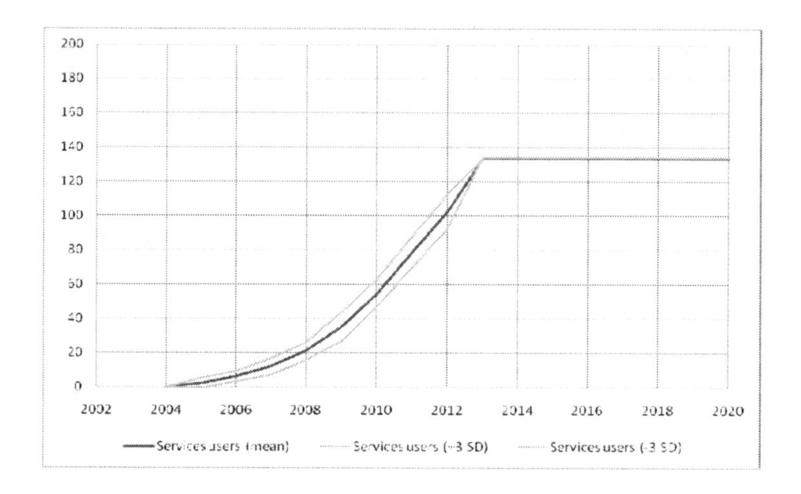

Fig. 6. Scenario 1 – No policies

Scenario 1: No policies. The use behavior (number of users of the service T-bollettino) without any governmental policies is shown in Figure 6. The use behavior has been obtained by the item of the questionnaire B21 "intention to use the T-bollettino service". The bold line represents the mean value, whereas the thin lines refer to minus and plus three times the standard deviation. In this scenario the growth is quite regular, slower during the first years and somehow quicker from 2009, without any singularities. The regime value is slightly below three quarters of the total number of DTV users and it is reached after the switch off in 2013.

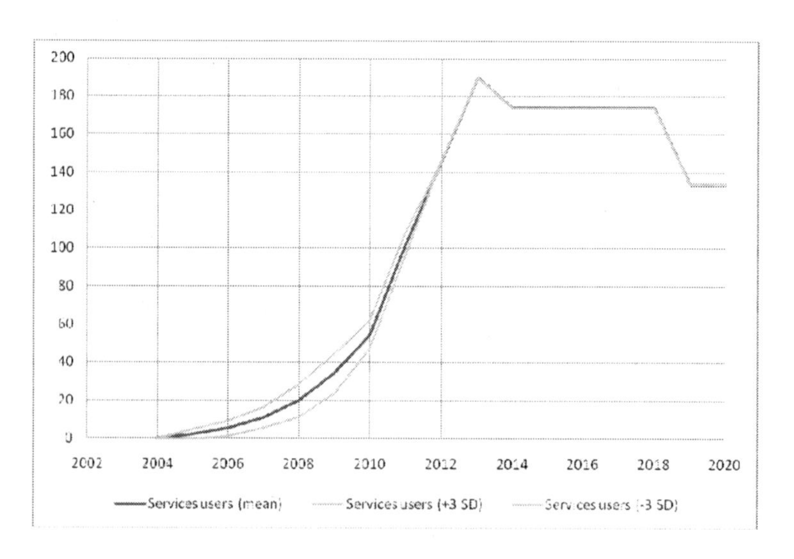

Fig. 7. Scenario 2 – Combination of policies PE (level 2) and FC (level 2)

Scenario 2: Combination of policies PE 2012-2013 (level 2) and FC 2011-2018 (level 2) Figure 7 shows a scenario where a combination of two different policies (PE and FC) is implemented in a realistic situation, where a communication campaign is financed by the national government for a couple of years around the switch off date with a strong level of intensity and a constant support to users through call centers and similar is provided for many years. The number of users who try the T-bollettino service goes to its upper limit in conjunction with the simultaneous action of the switch off, the campaign and the support, falling back to the regime value in two steps toward the end of the implementation of the policies.

3 Conclusions

The main aim of the study was to investigate user adoption and use behavior during the diffusion of DTV interactive services for citizens. The main result was a micro simulation model producing different strategic scenarios on citizens' use of DTV services depending on policies about DTV interactive service design, the informative campaign and as well as the support given to the residential population. The following main indications can be extracted from the scenarios:

- The adoption of a user centered design approach produces a significant increase of the frequency of use of the services during the implementation of the policy, in comparison to the scenarios in which no specific policies are adopted.
- Similarly, the realization of an intense communication campaign to showcase DTT services benefits to citizens produces a significant increase of the frequency of use during the implementation of the policy.
- The realization of a policy based on the provision of a strong user support (e.g., through a call centre) seems to produce smaller effects on use behavior in comparison to other policies.
- A combination of different policies conveys a significant increase of the service frequency of use during the implementation of the policies themselves.

These indications about the diffusion of interactive services are related to the limited context of the Italian pilot study. However, final results and achievements confirm that the adopted methodology is adequate for the study applied to the residential population. They also encourage future developments in the utilization of the UTAUT model. In fact, at this stage of the study about the usage of T-government services, the UTAUT key moderators (i.e. age, gender, experience, voluntariness of use) were not included in the process. On this respect a further analysis has been planned to aim at finding whether it is possible to incorporate in the model, in addition to the UTAUT key moderators, also the variables "income" and "education", either as new moderators or as new determinants.

Acknowledgements. These investigations were developed in the framework of the project "Monitor DTT" (Monitoring user experience with DTT) carried out by Fondazione Ugo Bordoni for the Italian Ministry of Economic Development – Department of Communication. The author wishes to warmly thank team members Michele Cornacchia, Enrico Nicolo' and Filomena Papa (FUB), and Stefano Livi (Sapienza University of Rome).

References

1. Davis, F.: User acceptance of information technology: system characteristics, user perceptions and behavioral impacts. Int. J. Man-Machine Studies, AP 38, 475–487 (1993)
2. Papa, F., Nicolo', E., Cornacchia, M., Sapio, B., Livi, S., Turk, T.: Adoption and use of digital tv services for citizens. In: Proceedings of the COST 298 Conference "The Good, the Bad and the Challenging", Copenhagen, pp. 161–171 (2009)
3. Papa, F., Nicolo', E., Livi, S., Sapio, B., Cornacchia, M.: Factors Affecting the Usage of Payment Services through Digital Television in Italy. In: Proceedings of EuroITV 2010, Tampere, June 9-11 (2010)
4. Papa, F., Spedaletti, S.: Broadband cellular radio telecommunication technologies in distance learning: a human factors field study. Personal and Ubiquitous Computing 5, 231–242 (2001)
5. Sapio, B., Turk, T., Cornacchia, M., Papa, F., Nicolò, E., Livi, S.: Building scenarios of digital television adoption: a pilot study. Technology Analysis & Strategic Management 22(1), 43–63 (2010)
6. Venkatesh, V., Davis, F.D.: A theoretical extension of the technology acceptance model: Four longitudinal field studies. Management Science 46(2), 186–204 (2000)
7. Venkatesh, V., Morris, M.G., Davis, G.B., Davis, F.D.: User Acceptance of Information Technology: Toward a Unified View. MIS Quarterly 27(3), 425–478 (2003)

Introduction to Economics and Governance of Future Networks

Bruno Tuffin

INRIA Rennes Bretagne-Atlantique, Campus de Beaulieu, France
bruno.tuffin@inria.fr

With the Internet evolution, and the development of next generation networks, there are natural questions about how the network will be managed and if/how public intervention will be made possible to improve its mode of functioning, as well as what will be or should be the business model representing the economic relations between all involved actors. Those issues are related to the so-called Internet governance, defined as the development of principle rules to shape the evolution and use of the Internet. Those questions arise especially because of the evolution from an academic to an interconnection of private networks and actors, and the emergence of new technologies that have to be fully deployed. The decentralized nature of the Internet of next generation networks in general, makes it also difficult to lead to a (socially) acceptable situation, with economic actors often trying to behave only towards what suits them the best.

This chapter will partially answer those questions by starting from the governance and public intervention issue and continuing with key aspects of business models attached to next generation networks.

Section 1 on "Public Intervention in the Deployment of NGNs" by J. L. Gómez-Barroso and C. Feijóo, describes the public intervention in the deployment of next generation networks, explaining why and how a direct intervention is currently implemented, making the connection with the universality of access principle. More specifically, it also discusses the related issue of public investments promotion in next generation networks, describing the guidelines of the European Commission on that problem. The idea is to define when public aid is allowed or not, depending mainly on whether broadband is already available or not and on the number of available providers.

The work on "Public Private Partnerships and Next Generation Networks" in Section 2, by I. Williams and M. Falch discusses the much related public-private partnership in the development of next generation access infrastructure, which can be required for an efficient network, and network development, in certain areas. It is especially discussed how can be applied in sub-Saharan Africa.

Also related to this public/private relationship, Section 3 on "Internet Governance and Economics of Network Neutrality" by P. Maillé, P. Reichl and B. Tuffin, introduces the reader to the network neutrality debate, which discusses if access providers could charge or treat differently packets depending on their source and/or their type, a behavior opposite to the Internet initial philosophy. This issue is at the

A.M. Hadjiantonis and B. Stiller (Eds.): Telecommunication Economics, LNCS 7216, pp. 89–90, 2012.

heart of the network governance, to understand if neutrality regulation rules have to be implemented, and at the heart of the next generation networks business model, by defining the possible economic relations between content and access providers, among others.

From Section 4 to Section 7, the chapter focuses on business models related to the deployment and evolution of next generation networks.

Section 4 on "Green Networks and Green Tariffs as Driven by User Service Demand" by L.-F. Pau describes an approximate model of energy consumption, capital expenditures and operating costs of a green 3G/LTE wireless network, i.e., a network minimizing energy consumption and CO2 emissions; such modeling is necessary to limit energy costs. The model characterizes the energy consumption due to the architecture, the type of traffic, and the services requested by users. The analysis is extended to renewable energy sources and allows determining jointly green tariffs and the operator profit impact.

The work on "Environmental Impact of ICT on the Transport Sector" in Section 5 by M. Falch discusses the environmental impact of ICT on the transportation system. It is in this sense related to the green ICT issue introduced in previous section because it searches to quantify the energy savings due to ICT thanks to telecommuting, teleshopping, or teleconferencing.

In Section 6 on "Cost-efficient NGN Rollout" by S. Verbrugge, J. Van Ooteghem, K. Casier, M. Van der Wee, and M. Tahon, business models for the deployment of optic fiber networks are introduced. Because fiber-to-the-home is deployed at a smaller speed than expected, the section describes the evolution in several European countries, identifies the reasons of the differences, proposes a holistic approach to improve the business, and presents a game-theoretic model to evaluate the impact of municipality investments on the market.

Finally, the work on "Software Business in the Telecommunications Sector" in Section 7 by L. Frank, E. Luoma, O. Mazhelis, M. Pulkkinen, and P. Tyrväinen, is about the software business model in the next generation networks. It introduces the evolution of the software market, the current trend for outsourcing, its interest with software seen as a service, a discussion on open source software adoption, and of future scenarios.

Public Intervention in the Deployment of NGNs

José Luis Gómez-Barroso[1] and Claudio Feijóo[2]

[1] UNED – Universidad Nacional de Educación a Distancia, Spain
[2] Universidad Politécnica de Madrid, Spain
jlgomez@cee.uned.es, cfeijoo@cedint.upm.es

Abstract. The deployment of NGNs (Next Generation Networks) has led to a change for the public role in the telecommunications industry. Now the public support is seen as not only responsible for the regulatory framework where market forces thrive, but as the leading force behind infrastructure deployment and innovative boosting measures. This section reviews the ways in which involvement of public administrations in the deployment of NGNs is happening. Subsequently, it focuses on the analysis of the European case and particularly on the Guidelines for the application of State aid rules *in relation to rapid deployment of broadband networks* that the European Commission published on Sept. 2009.

Keywords: telecommunications, next generation networks, public intervention, infrastructure deployment, public funds, State aid, broadband.

1 Introduction

Public activity in the telecommunications sector in the 21st century has experienced reorientations of certain importance, when compared with what was considered the norm at the end of the previous century. Aspirations for the benefits of an information society, evidence of different types of digital divides, and the need to move to a knowledge-based economy to maintain economic growth[1] and quality of life all influence a renewed interest in public involvement in the domain.

Programs intended to facilitate the development of the information society have been used as an umbrella for most of these resuscitated initiatives. In all of these plans, access to advanced telecommunications networks is mainly left to market forces [1]. This is the case particularly for "traditional" broadband services. However, problems related to the deployment of next-generation access (NGA) networks are considered conceptually distinct. These are problems that many can be considered as insurmountable without public involvement, which puts stress on policy-makers.

[1] On the occasion of the presentation of the Connecting Europe Facility (see next footnote), Neelie Kroes, Vice-President of the European Commission responsible for the Digital Agenda, affirmed that "over just ten years, the right broadband development could give Europe over one trillion Euro in additional economic activity, and create millions of jobs. An increase in broadband penetration of 10 percentage points would increase Europe's annual GDP growth by between 0.9 and 1.5 %".

A.M. Hadjiantonis and B. Stiller (Eds.): Telecommunication Economics, LNCS 7216, pp. 91–99, 2012.

As a consequence, there have been frequent allocations of public funds for the construction of infrastructure generally in places where it is believed that the market will not suffice, although similar projects can even be found in places where telecommunication operators maintain the availability of broadband connections under normal market conditions[2].

One of the most significant characteristics of this public "comeback" in the deployment of telecommunication networks is the fact that the decisions are being made, in most cases, by public entities at a "lower" level than that of the central government: regional authorities and especially local or town councils, who feel closer to the immediate needs of citizens [2, 3]. Indeed, many particular projects have been carried through autonomously without support from any national plans.

From a technological standpoint, the commercial maturity of some wireless standards, along with the resulting cost reductions, has actually multiplied the number of cases of broadband public networks using the radio spectrum [2, 3]. However, wireless technologies are not the only ones being used. There exist many examples of public administrations having driven forward or contributed to the deployment of cable or fiber [4]. Indeed, the scale is tilting towards wired technologies as NGA networks are introduced into the formula. Particularly illustrative to this respect are the *Guidelines for the application of State aid rules in relation to the rapid deployment of broadband networks* that the European Commission published on September 30, 2009, which will be analyzed in Section 3[3].

In terms of practical matters, there are no common rules governing how to sponsor the construction of these networks or manage them once they are completed. The strategies being deployed are diverse and basically depend on the specific circumstances of each case [5]. Importantly, many of the current public interventions involve private initiatives, including public-private partnerships [6].

Besides direct intervention in the deployment of infrastructures, there is a whole array of other policy instruments that can support the development of NGA networks. Most of plans also include support for adoption strategies, where public

---

[2] At the time of editing this contribution, the European Commission itself announced the investment of €9.2 billion from 2014 to 2020 on pan-European projects "to give EU citizens and businesses access to high-speed broadband networks and the services that run on them" At least €7 billion would be available for investment in high-speed broadband infrastructure. The funding, part of the proposed Connecting Europe Facility, would take the form of both equity and debt instruments and grants, "complementing" private investment and public money at local, regional and national level and EU structural or cohesion funds.

[3] Indeed, as stated in next section, a specific definition of the NGA is given in the Guidelines: they are cabled networks. However, in a Communication of 2006 ("Bridging the broadband gap") the European Commission itself had written: "The optimal mix of technologies depends on the characteristics of each particular location. The cost of technologies varies according to the number of potential users, the distance of the dwellings from the point of presence, and the presence of the backhaul. A scarcely-populated isolated area may be better off with a wireless solution and a small town with a wireline solution (…) No specific technology option will offer the best connectivity in all situations. The optimum is often achieved by a combination of technologies and solutions".

administrations, in addition to taking on direct and arbitrator roles, use indirect intervention in the electronic communications market, encouraging demand and improving the conditions for the supply activity of private players. For instance, the role of local governments is manifold. The most well-known and analyzed among these "municipal interventions" is the deployment of municipal broadband networks [7, 8]. But they also can contribute to the collection of information to create a precise map of available infrastructures and the uptake of interested users. They can help aggregate local demand, and they know local topography and therefore may contribute to the determination of the optimal technology mix or may synchronize civic works. More, they can also develop local public services, take part in pilot projects or living labs to explore new technologies and new interests of users, or even support the rollout of next generation infrastructures.

In this regard, it is also worthwhile to remember that many digital policies (eInclusion, eHealth, eGovernment, eParticipation) share the same basic axiom of preventing exclusion, which was originally part of the universal service approach, although a more balanced view of access/supply and adoption/demand is now prevalent. Therefore, the relationship between public and private can (and should) be put in a wider, more social and citizen-centric context. Understanding these new relationships is key to the development of NGA because they are no longer isolated from issues of general socio-economic welfare: NGA infrastructures are not an end in themselves; they are usually deemed an intermediate step to increase and widen innovation, to create new opportunities for employment and, to contribute broadly to productivity, economic growth, social equity and sustainability.

2 Promoting Investments in NGNs

On September 30, 2009, the European Commission published Guidelines for the application of State aid rules *in relation to the rapid deployment of broadband networks*. These Guidelines are applicable to two markets that are connected, although different: "traditional" or "first generation" broadband networks (both terms are used in the Guidelines) and next-generation access (NGA) networks[4]. In the next sections, these Guidelines are analyzed in order to clarify the conditions required by the European Commission to accelerate the deployment of NGA networks[5]. Despite the fact that the title of these Guidelines specifically refers to "the application of the rules on State aids", the truth is that their content describes in detail different procedures for public funding, which are not exclusively limited to State aid in a strict sense.

---

[4] According to the Guidelines, NGA networks are "wired access networks which consist wholly or in part of optical elements and which are capable of delivering broadband access services with enhanced characteristics (such as higher throughput) as compared to those provided over existing copper networks".

[5] On April 2011, the Commission launched a public consultation on the revision of the Guidelines. The Guidelines foresee their revision by September 2012 "on the basis of important market, technological and regulatory developments".

2.1 Use of Public Funds That Do Not Constitute State Aid

According to Article 107(1) of the Treaty on the functioning of the European Union, State aid is "any aid granted by a Member State or through State resources in any form whatsoever which distorts or threatens to distort competition by favoring certain undertakings or the production of certain goods shall, in so far as it affects trade between Member States, be incompatible with the internal market" (see [9, 10]). It follows that in order for a measure to qualify as State aid, the following cumulative conditions have to be met: (a) the measure has to be granted out of State resources; (b) it has to confer an economic advantage to undertakings; (c) the advantage has to be selective and distort or threaten to distort competition; (d) the measure has to affect intra-Community trade.

Aid outside the scope of Article 107(1) (i.e. aid that does not have all four conditions) includes State support of the roll-out of broadband by way of an equity participation or capital injection into a company that is to carry out the project, the case where Member States consider that the provision of a broadband network should be regarded as a service of a general economic interest (SGEI), or general measures available to all economic undertakings in all parts of the Member State.

2.1.1 Investment

Following the principle of equal treatment (and according to the case-law of the Court of Justice of the European Communities), the Guidelines establish that capital placed by the State, directly or indirectly, at the disposal of an undertaking "in circumstances which correspond to normal market conditions" cannot be regarded as State aid [11].

The conformity of a public investment with the application of the "market economy investor principle" has to be demonstrated by the following means:

- The existence of a sound business plan showing an adequate return on investment.
- Significant participation of private investors who take part in the project and assume the commercial risk linked to the investment under the same terms and conditions as the public investor.

2.1.2 Imposition of a Service of a General Economic Interest (SGEI) Mission

The determination of the nature and scope of an SGEI mission falls within the competence and discretionary powers of Member States. However, such competence is "neither unlimited nor can it be exercised arbitrarily", which, in the case of broadband networks, means that "in areas where private investors have already invested in a broadband network infrastructure (or are in the process of expanding further their network infrastructure) and are already providing competitive broadband services with an adequate broadband coverage, setting up a parallel competitive and publicly-funded broadband infrastructure should not be considered as an SGEI".

However, public service compensation may be granted to an undertaking entrusted with the operation of an SGEI on the condition that it can be demonstrated that private investors may not be in a position to provide adequate broadband coverage to all citizens or users in the "near future" (understood as referring to a period of 3 years).

According to the case-law of the Court, State funding for the provision of an SGEI may not be regarded as State aid, provided that four conditions are met. Those conditions are commonly referred to as "the Altmark criteria" [12, 13]:

a) the beneficiary of a State funding mechanism for an SGEI must be formally entrusted with the provision and discharge of an SGEI, the obligations of which must be clearly defined;
b) the parameters for calculating the compensation must be established beforehand in an objective and transparent manner;
c) the compensation cannot exceed what is necessary to cover all or part of the costs incurred in the discharge of the SGEI, taking into account the relevant receipts and a reasonable profit; and
d) where the beneficiary is not chosen pursuant to a public procurement procedure, the level of compensation granted must be determined on the basis of an analysis of the costs that a typical "well-run" undertaking would have incurred.

The first of the criteria demands "clearly defined obligations". In the deployment of broadband networks, there are at least two obligations: to connect all citizens and businesses, and to make the network available for all interested operators, allowing effective competition at the retail level (the publicly funded network should be a passive, neutral and open-access infrastructure).

2.1.3 Other Alternatives

Member States may decide to adopt a set of measures to accelerate the NGA investment cycle and, thus, encourage investors to advance their investment plans. Specifically, this consists of carrying out civil works. Indeed, a large part of the cost of deploying telecommunication networks (particularly fiber networks) is incurred in digging, laying down cables or installing in-house wiring. In this respect, Member States may decide, for instance, to undertake some works such as digging of the public domain, or construction of ducts. For this practice to be admissible, such civil works should not be "industry or sector specific" but should be open to all potential users and not just electronic communications operators (e.g., electricity, gas or water utilities).

The Guidelines also mention options such as easing the acquisition process of rights of ways or demanding that network operators coordinate their civil works and/or share part of their infrastructure. Note, however, that these are regulatory measures that do not require any public funds to be paid out.

2.2 State Aid

A part of the State aids granted for the deployment of broadband networks could be considered to be included in the exception stated in item a) of article 107, section 3, of the Treaty on the functioning of the European Union ("aid to promote the economic development of areas where the standard of living is abnormally low or where there is serious underemployment"). However, it is item c) of this section that makes generally acceptable the aid granted by the States for these projects ("aid to facilitate the development of certain economic activities or of certain economic areas, where such aid does not adversely affect trading conditions to an extent contrary to the common interest").

In assessing whether any specific measure can be deemed compatible with the common market, the Guidelines for the broadband aids set out a background defined through a "colored area" analysis:

- "White" areas are those where broadband is currently not available and where there are no plans by private investors to roll out such an infrastructure in the near future.

  The term "in the near future" should be understood as referring to a period of three years. In this regard, investment efforts planned by private investors should not guarantee that significant progress in terms of coverage will be made within the three-year period (with completion of the planned investment foreseen within a reasonable time frame thereafter).

  In the Guidelines it is recognized that the Commission has taken *an overwhelmingly favorable view* towards State measures for broadband deployment for underserved, typically rural, areas.
- "Grey" areas are those where only one broadband network operator is present.

  In these situations, State support for the deployment of broadband networks calls for "a more detailed analysis and careful compatibility assessment" (as monopoly provision "may affect the quality of service or the price at which services are offered to the citizens", but, on the other hand, subsidies for the construction of an alternative network can distort market dynamics).

  If more than one network will be deployed in the "near future" (the coming three years), such an area should, in principle, be considered black (see below).
- "Black" areas are geographical zones where at least two broadband network providers are present and where broadband services are provided under competitive conditions[6].

In black NGA areas, the wording is clear: "no need for State intervention".

It is important to underline that the case of white NGA areas were, at the moment where the Guidelines were published, the rule in the European Union. Note also that part of the areas that are grey and even black for the traditional broadband are white for the NGA networks, as there has been no deployment in these areas. For cases where there is a "change of color", the Guidelines establish "compatibility" rules:

- In traditional grey areas (but white NGA areas), the grant of aid for NGA networks is subject to the demonstration by the Member State concerned that "the planned NGA network is not or would not be sufficient to satisfy the needs of citizens and business users in the areas in question" (also taking into account a possible future upgrade).

---

[6] Therefore, facilities-based competition exists. There could also exist competition at the level of services in grey areas, should the single network be open to other operators.

- In black (basic broadband) areas no further State intervention should *in principle* (same clarification again) be necessary, as "existing network operators should have the incentives to upgrade their current traditional broadband networks to very fast NGA networks to which they could migrate their existing customers".

  Member States can rebut such an argument "by showing that existing basic broadband operators do not plan to invest in NGA networks in the coming three years by demonstrating for instance that the historical pattern of the investments made by the existing network investors over the last years in upgrading their broadband infrastructures to provide higher speeds in response to users' demands was not satisfactory".

It must be underlined that an additional qualification is made for projects to be authorized; they will be approved provided that there are no less distortive means (including *ex ante* regulation) to reach the stated goals.

If this requirement is met and a balancing test of the project concludes that the overall evaluation of it is positive, a number of fundamental conditions must be complied with to "minimize the State aid involved and the potential distortions of competition". These conditions are detailed mapping exercise and coverage analysis, open tender process, priority to the most economically advantageous offer, technological neutrality, re-use of existing infrastructure, mandated wholesale open access (for at least seven years)[7], benchmarking exercises (with average published or regulated wholesale prices that prevail in other comparable areas of the country) and claw-back mechanisms (to ensure that the selected bidder is not overcompensated if the demand for broadband in the target area grows beyond anticipated levels).

3 Summary and Conclusions

The advent of the new socio-economic paradigm represented by the information society presents countries with risks and opportunities. The support for information technologies and service industries appears to be a sure path for future economic development. Indeed, this may be the only way for countries with narrow domestic markets or limited physical resources to prosperity. This general remark is particularly true in times of crisis, when countries are desperately searching for ways to return the economy to its growth path.

In this scenario, NGNs are regarded as a necessary tool for prosperity. This justifies a much more active intervention in the telecommunications industry's activity, and even allocating public funding for the deployment of networks [14]. Conceptually, this is nothing particularly new. Generalized access to telecommunication services has been, in general and regardless of the degree of success, an objective of every

[7] Whatever the type of the NGA network technical architecture, it should support effective and full unbundling and satisfy all different types of network access that operators may seek (including but not limited to access to ducts, fibre and the so-called "bitstream access"); in setting the conditions for wholesale network access, Member States should consult the relevant NRA.

government during the last century [15]. This suggests that the advantages of a massive connection to telecommunication services have been understood regardless of the political option in power. What can indeed be considered a novelty as regards the general rule during the last decade, is the depth of the intervention. Following the complete liberalization of the markets it seemed to have been accepted that the activity of the private agents could only be complemented exceptionally to correct the "failures" of the offer in networks that were then almost universal [16]. Indeed, the contribution of public funds for the deployment of new networks did not seem foreseeable only a few years earlier, at least in developed countries.

However, the presently reinforced public intervention should not be done behind or regardless of the market. It is worthy of note that, in spite of the supply-side focus of many plans, uncertainties about the demand subsist, thus arguing for more integrated approaches and not just the simple deployment of NGNs. The, up to now, unremitting impact of the economic crisis in the capacity of expenditure of public administrations is leading also to look for the best conditions for the public investment.

References

1. Gómez-Barroso, J.L., Feijóo, C., Karnitis, E.: The European policy for the development of an information society: the right path? Journal of Common Market Studies 46(4), 787–825 (2008)
2. Strover, S., Mun, S.H.: Wireless broadband, communities, and the shape of things to come. Government Information Quarterly 23(3-4), 348–358 (2006)
3. Troulos, C., Maglaris, V.: Factors determining municipal broadband strategies across Europe. Telecommunications Policy 35(9-10), 842–856 (2011)
4. Ehrler, M., Ruhle, E.O., Brusic, I., Reichl, W.: Deployment of fibre optic networks within the framework of public-private partnerships projects. International Journal of Management and Network Economics 1(3), 308–326 (2009)
5. Ragoobar, T., Whalley, J., Harle, D.: Public and private intervention for next-generation access deployment: Possibilities for three European countries. Telecommunications Policy 35(9-10), 827–841 (2011)
6. Falch, M., Henten, A.: Public private partnerships as a tool for stimulating investments in broadband. Telecommunications Policy 34(9), 496–504 (2010)
7. Mandviwalla, M., Jain, A., Fesenmaier, J., Smith, J., Weinberg, P., Meyers, G.: Municipal broadband wireless network. Communications of the ACM 51(2), 72–80 (2008)
8. Goth, G.: Municipal wireless gets new emphasis – Rethinking the infrastructure cloud. IEEE Internet Computing 13(6), 6–8 (2009)
9. Friederiszick, H.W., Röller, L.H., Verouden, V.: European state aid control. An economic framework. In: Buccirossi, P. (ed.) Handbook of Antitrust Economics, pp. 625–669. MIT Press, Cambridge (2008)
10. Kurcz, B., Vallindas, D.: Can general measures be... selective? Some thoughts on the interpretation of a state aid definition. Common Market Law Review 45(1), 159–182 (2008)

11. Parish, M.: On the private investor principle. European Law Review (1), 70–89 (2003)
12. Renzulli, A.: Services of general economic interest: The post–'Altmark scenario'. European Public Law 14(3), 399–432 (2008)
13. Klages, R., Alemanno, A., Falcone, N., Petit, N., Ballester, R.: Services économiques d'intérêt général (arrêt 'Altmark Trans GmbH'); Revue du Droit de l'Union Européenne, No. 2003-3, pp. 755–759 (2003)
14. Cave, M., Martin, I.: Motives and means for public investment in nationwide next generation networks. Telecommunications Policy 34(9), 505–512 (2010)
15. Gómez-Barroso, J.L., Feijóo, C.: A conceptual framework for public-private interplay in the telecommunications sector. Telecommunications Policy 34(9), 487–495 (2010)
16. Gómez-Barroso, J.L., Pérez-Martínez, J.: Public intervention in the access to advanced telecommunication services: assessing its theoretical economic basis. Government Information Quarterly 22(3), 489–504 (2005)

Public Private Partnerships and Next Generation Networks

Idongesit Williams and Morten Falch

CMI, Aalborg University, Copenhagen, Denmark
idong@plan.aau.dk, falch@cmi.aau.dk

Abstract. This section discusses the possible role of Public Private Partnership (PPP) in the development of infrastructure for Next Generation Networks. The focus is on the developing countries perspective. Case studies describe the African approach in using PPP to fund telecom infrastructure development in rural areas.

Keywords: Public Private Partnership, Next Generation Networks, Governance, Universal Access, Developing country.

1 Introduction

As Information and Communications Technologies move towards Next Generation Networks (NGNs), broadband has moved from "just one of those telecommunication network technologies" to a national resource as identified by the secretary general of the ITU [19]. This has heightened the attention of Governments and other policy bodies on the potentials of broadband technology to the economy, as well as the social lives of their citizens. Governments around the world are looking for efficient ways for partnering with the private sector to develop broadband and attain their universal service objectives as well. Hence in the development of NGNs, various forms of Public Private Partnerships (PPP) will play an important role in the development of Next Generation Access (NGA) infrastructure. PPP may not necessarily be implemented as it is today, but the market situation and the socio-economic realities that may act as an impediment to an effective working of the market in certain areas will justify the use of PPP to initiate a market based development.

Still more capacity can be offered by wireless services and wireless network solutions are becoming viable alternatives to wired network infrastructures. This has led to a reduction in costs of diffusion and adoption of ICTs. Still there will be some remote areas, which are difficult to reach, and where provision of broadband connectivity will remain unprofitable from a commercial point of view. Here PPP can be deployed in order to extend connectivity to these remote areas also. Especially in developing countries this is a pertinent problem. Here, rural areas are often poorly equipped with infrastructure facilities and the customer base is weak due to low income levels and a sparse population. This section of the book takes a look at some of the current PPP approaches towards development of broadband and other network services, and discusses how these can be applied in sub-Saharan Africa.

A.M. Hadjiantonis and B. Stiller (Eds.): Telecommunication Economics, LNCS 7216, pp. 100–107, 2012.

2 The Concept of PPP

There are many different definitions of the concept of PPP. PPP has been described as "a cooperative venture between the public and the private sectors built on the expertise of each partner, that best meets clearly defined public needs through the appropriation of resources risks and awards" (CCPPP). This cooperation, of course, varies in its financial, organizational and implementation design. According to the EU Green Paper on Public Private Partnership [2], the private actors should do the implementation, must take part in the funding and should assume at least a part of the financial risk, while it is up to the public actor to define objectives and monitoring.

For a number of years, PPP was seen as a way to involve private companies in publicly initiated activities [15]. This could, for instance, be outsourcing of care for the elderly to private businesses operating under public supervision or licenses. The liberalization of telecommunications can also, to some extent, be seen in this view with commercial operators guided by public policy demands on the sector with respect to universal service requirements and price regulations. However, the present discourse on PPP in the telecommunications area is turning the other way round. Now, the issue is the involvement of public money in providing rural access to privately owned broadband infrastructure [4].

This has led to the present discourse on PPP in the telecommunications where public money is being spent as a cushion to private financing in order to help the private sector achieve the universal access objectives set by public authorities. This new financing option has not in any way affected the initial PPP objectives from both parties. Each party expects trust, openness, fairness and mutual respect from the other party [10]. The expectations of the two parties remain the same even as the financing options and spectrum of PPP possibilities continue to evolve. The public agency expects an improvement in program performance, cost efficiencies, better provision of services and appropriate allocation of risks and responsibilities [14]. The private sector expects a better investment potential that will lead to reasonable profits. This will in turn open opportunities for them to expand their businesses [17]. However, this new push by the public in some cases has led to the development of a new spectrum of concepts of PPP. The implementation of PPP may take different avenues but the basic idea of cooperation between the public and private sector remain the same irrespective of how the financial obligations are designed.

There are other ways in which public telecommunication agencies can fund Universal Access and Service of NGNs. These will continue to be used of course to fund infrastructure development of NGNs as identified by [3]. Some of them are:

- Universal Assess and Service funds (UASF)
- Public funding programs and investments
- Investment by private companies other than telecom operators
- Funding through Non-Profit Organizations

However, PPP has some advantages that make the concept appealing and beneficial to the development of NGN infrastructure in Sub-Saharan Africa and rural areas in the world. These includes, access to private finance for expanding services, clearer objectives, new ideas, flexibility, better planning, improved incentives for competitive tendering and greater value for money for public projects [13], [18].

Secondly, the four funding mechanisms listed above can be implemented via the vehicle of PPP. In Africa and many other parts of the world, UASFs provide subsidies to private companies for the development of telecommunication infrastructure in rural areas, underserved and non-served areas. There are such cases in Uganda, Mozambique, and a host of other African countries [8], [9]. In the case of Uganda, the USAF gets money from the World Bank and provides this money as seed capitals, a form of subsidy to the private sector to develop Universal Access projects. Public funded programs and investments can be a PPP if private sector is also contributing some form of capital to the project and investment by private companies can also be a PPP if a public agency is involved in some way as a collaborating partner in the project. Funding through NGO's may be a PPP if the NGO is either a public NGO or the NGO is partnering with some public agency to deliver the telecommunication infrastructure development.

The ability of PPP to be used in different ways makes it an interesting way of funding next generation networks in Africa, rural areas, non-served and underserved countries of the world.

3 Case Studies

So far, as a result of the problem of accessing remote and difficult to reach places, some countries are taking practical steps to develop a partnership with the private sector in the development of broadband infrastructure in these remote areas. Different forms of PPP financing options for telecommunication and other infrastructure all around the world include the Build-Operate-Transfer (BOT), Build-Transfer-Operate (BTO), Build-Own-Operate (BOO), Build-Own-Operate-Transfer (BOOT), Build-Lease-Transfer (BLT), Design-Build-Finance-Operate (DBFO) and Design Construct Manage Finance (DCMF) concepts. Most of these public-financing cases are very remote in the African telecommunication scene. Most of Africa embraced privatization and semi-privatization schemes during the liberalization era. However in other sectors of the economy in Africa like transportation etc., these forms of PPP have been used.

It is pertinent to point out that PPP today is regarded as a complement to a free market in the development of broadband infrastructure and the market. There will likely be fewer interventions in the development of Next Generation Access Networks in cities, metropolitan areas and commercially viable areas. As the family of fixed and mobile standards of Next Generation Networks mature, this will go a long way in enabling the diffusion and adoption of these services en masse. However, this maturity will not occur without the right business models and governance structures put in place by players in the next generation value chain and the public regulating bodies respectively.

It is still difficult beforehand to tell if a competition driven market alone will enable the desired objectives of universal access to the Next Generation Networks in a specific area, or if a PPP program will be an efficient way to achieve the desired policy objectives.

In Africa, the use of PPP to deploy infrastructure and thereby propping the market to spring up in the telecommunications sector can be seen in the user adoption side, the infrastructure development and Internet service development. NGNs are largely envisaged to be delivered via broadband (fixed and mobile) and IP networks. The infrastructure of these new networks may be cheaper or more expensive than the existing family of technologies. However, the approach of using PPP as a compliment to market mechanisms will likely continue as long as there are rural areas in Africa.

The four case studies below illustrate that various forms of PPP have been developed as a compliment to the competition fostered by liberalization, which gave way to privatization and open access of telecommunication service delivery and infrastructure. This has taken place at the continental and the national levels as well as at the regional level.

3.1 NEPAD Broadband Connectivity Programme – Continental Fiber Optics Deployment

In the African continent, there has been an ambitious move by New Partnership for African Development (NEPAD) to provide broadband connectivity in all the 54 African countries [12]. Basically, one might say Africa is getting ready for NGNs. The aim is to provide abundant bandwidth with easier connectivity at a reduced cost. All NEPAD head of states, which invariably are all the heads of States in Africa adopted the proposal, however not every country that is signatory to the project have embarked on proactive steps to ensure the possibility of this project. The development is divided into two areas, the NEPAD ICT broadband infrastructure network for Eastern and Southern Africa and the NEPAD ICT broadband infrastructure network for Central, West, and North Africa.

The Policy Principles developed for this project clearly point to the complementary role of PPP in relation to market development:

1. Non-discriminatory Open Access, whereby all Authorized Service Providers, in any country market, will access broadband connectivity on the same terms, including pricing. This will provide a level playing field for all Authorized Service Providers, increase competition, and thereby reduce bandwidth prices to the end-users.
2. Equitable Joint Ownership of the backbone infrastructure across the region.
3. Use of Special Purpose Vehicles (SPVs) to build, own and operate the Broadband ICT network
4. The Basic ICT Broadband Infrastructure should be viewed as a "public good".
5. Application of PPP in the development of the network.

Policies here as agreed by the policy makers covered both deployment of submarine cables as well as terrestrial networks [12]. NEPAD is the Public Agency involved in this project and the Special Purpose Vehicle is a private consortium involved in the project.

The concept of the Special Purpose Vehicle is encapsulated in the Kigali Protocol which encapsulates the policy principle listed above as well. The Special Purpose Vehicle owns, operate and maintain the broadband Network. The NEPAD submarine SPV are to be set up by African Telco and non-telecom investor. The African Investors have a stake of 45% while the International investors will have 25%. Funding for the PPP for the Broadband infrastructure Network for Eastern and Southern Africa were expected from international donors like Industrial Development Corporation (IDC), Development bank of Southern Africa (DBSA) and Pan African Infrastructure Development Fund (PAIDF).

This project is still being developed at the moment, but it gives a clear indication of the regional approach to the present broadband. This may be extended to the future towards next generation networks depending on the infrastructure needs. Other initiatives in the same area are GLO [6] and, SAT-3/WACS/SAFE [16].

3.2 PPP and User Adoption – Rwanda Mobile Telephone Adoption Scheme

In order to attract less affluent Africans, different African governments have taken some regulatory and PPP steps in addressing the problem. One of the major problems experienced in Africa was the affordability of mobile telephony handsets. This problem is likely to also continue into the Next Generation Network era of telecommunication networks. Hence various forms of PPP as well as regulation may still be relevant in propping demand for next generation networks by ensuring the affordability of Customer Premises Equipment (CPE).

On the supply side, marketing innovations played a major role in promoting the competition enabled by the liberalization. These innovations from the supply side triggered the adoption of ICT on the demand side. A clear example is mobile telephony. Mobile telephony, which has become a success in the region, was enabled by the introduction of the pre-paid mobile solutions as well as per-second billing in call termination charges. These drivers made mobile telephony attractive to the less affluent African who could now own a mobile telephone without the fear of incurring monthly charges in which the subscriber had to pay later as in the case of postpaid solutions. However, despite these successes, there were still issues of the high cost of mobile telephony handsets. Many governments had to intervene by the reduction of taxation as in the case of Ghana [11] and the use of subsidies to enable less affluent citizens acquire mobile telephony handsets as in the case of Rwanda. What makes the case of Rwanda unique is the partnership between the Government of Rwanda, MTN (a telecommunication operator) and the Rwanda Development Bank (BRD) [1].

3.3 PPP and Infrastructure Development – Uganda USAF Funding on Mobile Telephony Infrastructure

The liberalization brought in either full or partial privatization of incumbent monopolies. However Uganda has a good case where competition was used alongside a PPP effort in the development of telecommunication infrastructure in rural areas. A case study conducted by Econone paints the following scenario.

The liberalization of the Ugandan telecommunication sector in 1993 led to the entrance of two new operators in addition to the later partially privatized incumbent, Uganda Telecommunications Limited (UTL). MTN was licensed as the Second National Operator in 1998 and Celtel were licensed to operate in the south west of the country. MTN and UTL were required to provide full country coverage. However despite regulatory measures to ensure the universal access obligation granted to MTN and UTL, the Government established a body called the Rural Communications Development Fund in 2001. The fund is financed by the payment of 1% of gross revenue from the three operators. The policy specifies the provision of basic communication services to all sub-counties with at least 5000 inhabitants by the year 2005. Licenses are granted by the Uganda Communications Commission (UCC) to independent local operators in areas most difficult to serve. The financing plan was to grant subsidy assistance of US$1 for every US$1 of private investment in the ICT infrastructure. Financial institutions like the World Bank estimated that US$5.8 million of subsidy would be required to make the 2005 date feasible. The PPP in addition to the competitive market made possible the tele-density of Uganda to grow from 2 subscribers per 1000 people in 1998 to 32 subscribers per 1000 inhabitants in 2003 [5].

Other supporting examples: In Ghana, the Universal Access and Service Funds GIFEC is also engaging the PPP as a way of Common Telecommunications Sites facilities across Ghana. They collaborate with telecommunication companies in the award of Subsidies to telecommunication companies who are willing to develop infrastructure for co-location [7]. In Nigeria, The Universal Service Provision Fund provides subsidy to the building of GSM Base Transceiver Stations (BTS). There is a similar approach by the Nigerian Universal Access and service fund in delivering wholesale internet bandwidth to Community Communication Centers (CCC), Cybercafés, rural internet service providers etc. The essence of this PPP effort in Nigeria is to enable private operators to provide and operate broadband network in rural areas [20].

3.4 Internet Service Deployment – Egyptian Smart Village Project

In 2005 the Egyptian government embarked on the 'smart village' initiative concept. This project is a PPP between the Ministry of Information and Communication Technology (MCIT) and a private consortium. The MCIT provided 300 acres of land (20% of the cost) and the private investors financed the remaining 80%. Infrastructure provided in the 'smart village' is fiber optic network and a multi-source power supply. A related project is the free internet initiative. The aim of this project was to enable users across Egypt to access Internet at the cost of a local call with no additional subscription fee. The free internet initiative is based on an offloading/revenue sharing model. The Internet Service Providers (ISP) are allowed to co-locate their access equipment at telecom Egypt local exchanges and re-routed to the ISP data backbone. This resulted in the major offloading of the Telecom Egypt PSTN network. Revenue accrued was shared between telecom Egypt and service providers [9].

4 Conclusions

The realities of the deployment of NGNs in developed and developing countries will differ greatly depending on the approaches as seen in the cases stated. These approaches will depend on the economic realities of the country, the priority placed on the development of ICT over other more important economic needs, political will and the effects of globalization. These factors will play a crucial role in the public intervention of deploying NGNs.

At the moment, developing countries in Africa are yet to achieve their universal access targets on previous standards in contrast to their counterparts in the EU. The same can be said about some countries in Asia. Hence with these shortfalls, the same way the governments of developing countries saw the need to engage in PPP to compliment and enable competition of existing standards, this same approach will still be adopted and redefined over time using different approaches to reach areas where connectivity is still a challenge. PPP will enable network operators deploy NGNs at a cheaper cost than they would if they carried the burden all by themselves. The Governments may not necessarily provide subsidies, but they may provide incentive regulations or even adopt the various forms of PPP as a way of either sharing or lessening cost. As pointed out in the case of Rwanda, governments may still use PPP to reduce the cost of acquiring the NGN services. Although it is difficult to foresee the future, the relevance of PPP as a public intervention tool in the deployment of NGNs in the future can't be ruled out.

References

1. Balancing Act (November 2011), http://www.balancingact-africa.com/news/en/issue-no-390/telecoms/rwanda-s-government/en
2. Canadian Council for Public-Private partnerships (CCPPP), Available on the internet (November 2, 2011), http://www.pppcouncil.ca/aboutPPP_definition.asp.cited
3. European Commission. Green Paper on Public-Private Partnerships and Community Law on Public Contracts and Concessions. COM(2004) 327 final (2004)

4. Falch, M., Henten, A.: Investment dimensions in a Universal Service Perspective: Next Generation Networks, alternative funding mechanisms and Public-Private Partnerships. Info. 10(5/6), 33–45 (2008)
5. Falch, M., Henten, A.: Public Private Partnerships as tool for stimulating investment in broadband. Telecommunications Policy 34(9), 496–504 (2010)
6. Farlem, P.: Working together, Accessing Public-Private partnerships in Africa. OECD (2005), http://www.oecd.org/dataoecd/44/4/34867724.pdf (November 2011)
7. GLO1. (November 2011), http://www.gloworld.com/glo1.asp
8. GIFEC. (November 2011), http://gifec.gov.gh/index.php?option=com_content&view=article&id=77:telecommunication-operators&catid=38:projects&Itemid=248
9. ICT regulation toolkit. Public Private Partnerships (November 2011), http://www.ictregulationtoolkit.org/en/Section.3288.html
10. ICT. regulation toolkit. Public-Private Partnerships in the telecommunications and ICT sector (November 2011), http://www.ictregulationtoolkit.org/en/PracticeNote.3160.html
11. Jamali, D.: Success and failure mechanisms of Public Private Partnerships (PPPs) in developing countries. Insights from the Lebanese context. The International Journal of Public Sector Management 17(5), 414–430 (2004)
12. Klutse, F.: Mobile Phones now chapter in Ghana. IT Africa news (2008), http://www.itnewsafrica.com/2008/09/mobile-phones-now-cheaper-in-ghana (November 2011)
13. NEPAD East African Commission (November 2011), http://www.eafricacommission.org/projects/126/nepad-ict-broadband-infrastructure-network
14. Nijkamp, P., Van der Burch, M., Vidigni, G.: A comparative institutional evaluation of public private partnerships in Dutch urban land - use and revitalization projects. Urban studies 39(10), 1865–1880 (2002)
15. Pongsiri, N.: Regulation and Public Private Partnerships. The International Journal of Public Sector Management 15(6), 487–495 (2002)
16. Sadka, E.: Public-Private Partnerships: A Public Economics Perspective. IMF Working Paper, WP/06/77 (2006)
17. SAT-3/WASC/SAFE (November 2011), http://www.safe-sat3.co.za/
18. Scharle, P.: Public Private Partnership as a social game. Innovation 15(3), 227–252 (2002)
19. Spackman, M.: Public Private Partnerships: lessons from the British approach. Economic Systems 26, 283–301 (2002)
20. Toure, H.: Building our Networked Future based on Broadband. ITU (2010), http://www.itu.int/en/broadband/pages/overview.aspx
21. Universal Service Provision Fund. Universal Service Provision Annual Report 2009. Nigerian Communications Commission (2009) http://uspf.gov.ng/downloads.htm?p13_sectionid=2

Internet Governance and Economics of Network Neutrality

Patrick Maillé[1], Peter Reichl[2], and Bruno Tuffin[3]

[1] Télécom Bretagne, Cesson-Sévigné Cedex, France
[2] FTW Telecommunications Research Center Vienna, Austria
[3] INRIA Rennes Bretagne-Atlantique, Campus de Beaulieu, France
patrick.maille@telecom-bretagne.eu, reichl@ftw.at,
bruno.tuffin@inria.fr

Abstract. This section deals with the network neutrality debate to define the economic relationships between Internet access providers and (distant) content providers. We review the various definitions of what should be a neutral network, the outcome of all international consultations on the topic, and the arguments of network neutrality proponents and opponents.

Keywords: Network neutrality, Regulation, Content providers.

1 Introduction

The Internet has known an uncontested success during the past two decades, having evolved from an academic network interconnecting researchers who exchange information, to a commercial network accessed by all individuals, made of private access, content, and interconnected transit providers, and over which all type of commerce or content exchange can be performed. Complex relations between those often non-cooperative actors have raised discussions about whether or not and how current telecommunications businesses and technological models should be sustained.

A specific question that we want to discuss in this section is the economic relations between Content Providers (CP), i.e., organizations or individuals creating information, educational or entertainment content, and Internet Service Providers (ISP), i.e., the companies or organizations providing access to the network.

In the current Internet business model, ISPs charge both end-users and CPs directly connected to them, the economic relations between interconnected ISPs being generally through public peering or transit agreements. But a key fact is that ISPs do not charge CPs that are associated with other ISPs, and do not treat their traffic in a different way. In addition, there is a specific layers (or Tiers) structure in the relations between ISPs. The so-called « Tier 1 » operators are defined as operators that can reach every other network without any form of fee. Tier 1 operators peer with all other Tier 1 operators. Tier 2 operators on the other hand have to settle peering agreements with other networks, to reach some portions of the Internet. Usually, those operators purchase their traffic to a Tier-1 network rather than discussing with other Tier 2 operators (because it is easier to implement).

A.M. Hadjiantonis and B. Stiller (Eds.): Telecommunication Economics, LNCS 7216, pp. 108–116, 2012.

But this business model has been questioned, especially because there is now an important asymmetry in the traffic volume exchanges between ISPs, mainly due to some prominent and resource consuming content providers. The typical example is YouTube (owned by Google), accessed by all users while hosted by a single Tier 1 ISP, and whose traffic constitutes now a non-negligible part of the whole Internet traffic. Moreover, while transit revenues made by ISPs are constantly decreasing, content providers revenues are increasing. As an illustration, it is predicted (by Jupiter Research) for 2012 an amount of $34.5 billion in online advertising spending in the USA only, while the transit costs are believed to be under 1$ per Mbps by 2014. This issue has been the reason why Ed Whitacre, CEO of AT&T, launched the debate about revenue sharing and the business model of the Internet in 2005 [8]. Basically, Ed Whitacre claimed that distant CPs use the AT&T network for free in order to reach end users, and that it could not be economically sustained: they should be charged by ISPs they are not directly connected to, especially because of the decreasing transit prices. Indeed, the ISPs do not have any incentive to invest into a network architecture, by improving the actual one or by implementing a new one. Another management possibility to differentiate CPs, not related to charging, has been implemented by the USA provider Comcast in 2007, which blocked P2P applications such as BitTorrent, under the argument that P2P content is mostly illegal.

All this has led to a strong protest from CPs who are losing accessibility or quality of service, and from user associations claiming that it is an infringement of freedom of speech. This raised the so-called *Network Neutrality Debate*, bringing questions at the same time at the economical, legal, technical, and political levels. The purpose of this section is to review the proposed definitions of network neutrality, a notion actually not formalized, to summarize the outcomes of the international consultations held very recently, and to describe and discuss the arguments of both proponents and opponents of neutrality to allow the reader to make her own conclusions.

It is important to remark that even governments somewhat see the position of CPs as an issue, in accordance with ISPs complaint. In France for instance, there is a worry about the online advertising revenue not being taxed as usually done, and a proposition on that matter has been released (see [13]). Between 10 and 20 million Euros could be collected in taxes from the USA providers such as Google, Microsoft, Yahoo and Facebook.

2 Network Neutrality Definitions

Maybe surprisingly, there is no clear and generally admitted definition of network neutrality. A network is often said to be weakly neutral if it prohibits user discrimination [11,12], which means it does not charge users differently for the same service, but can differentiate between applications (example: video having a priority with respect to email because of more stringent QoS requirements). It is said to be strongly neutral if in addition it prohibits any service differentiation that is if it does not allow managing packets differently.

A tentative of « official » definition of neutrality was made in 2005 in the USA by the Federal Communications Commission (FCC) stating that network neutrality could be summarized into four items [3]: (1) no content access can be denied to users (2) users are free to use the applications of their choice (3) they can also use any type of terminal, provided it does not harm the network (4) they can competitively select the access and content providers of their choice. This corresponds exactly to the well-known proposition/definition has been described by Professors Timothy Wu and Lawrence Lessig; summarized by Tim Wu, "Network neutrality is best defined as a network design principle. The idea is that a maximally useful public information network aspires to treat all content, sites, and platforms equally."

3 International Consultations

The debate has been so strong, and the potential impact of a regulation or non-regulation so important that governments have held public consultations. This has started in the USA, but and also recently in France (by ARCEP, the French regulator), UK (by Ofcom) and by the European Commission (EC) in 2010, after Norway (by the Post and Telecommunications Authority, NPT) and Sweden (by the post and Telecom Agency, PTS) in 2009. In that same year, the FCC has imposed neutrality regulatory rules, a decision that has been appealed by an operator, Verizon. The goal of those consultations was to (re)design an Internet that suits the interests of all actors (consumers, broadband service providers, entrepreneurs, investors, and businesses of all sizes) thanks to the most appropriate trade-off. Even if everyone was allowed to respond to the consultation, it has mainly been answered by specialists such as computer scientists, industry lobbies and non-governmental organizations. The participation was different across the countries, and when individuals were responding, they were largely defending neutrality.

One of the most important outcomes/recommendations is that traffic management has to be *transparent*, in the sense that operators have to clearly explain what they are implementing. This is also considered as being a manner to avoid anti-competitive behaviors from operators and to control if a minimal best-effort Internet is in place. Transparency has been a consensus among all industrial telecommunication actors. ARCEP proposes that operators provide all clear and relevant information on access, QoS, and limitations (if any) [1,10].

Another key idea is to protect customers' rights to access content, which cannot be forbidden, i.e., the network should be *open*, which is defined as allowing "end users in general to access and distribute information or run applications of their choice" [2].

Another aspect is quality of service (QoS). A result of consultations is the demand to define a minimum level of QoS (which has to be defined), for a type of *universal access* principle. Weak neutrality and service degradation would be allowed if such a minimum level is reached. No regulator seems to have something against discrimination if a minimum QoS is guaranteed. It is even considered that a total lack of traffic management could result in a lower perceived QoS for users. In this sense, being neutral and open does not necessarily mean that all bits have to be considered equal. The FCC advocates for a "reasonable network management" vaguely defined as " (a) reasonable practices employed by a provider of broadband Internet access service to (i) reduce or mitigate the effects of congestion on its network or to address quality-of-service concerns; (ii) address traffic that is unwanted by users or harmful; (iii) prevent the transfer of unlawful content; or (iv) prevent the unlawful transfer of content; and (b) other reasonable network management practices" [4]. This could be realized as soon as transparency is fulfilled.

All those aspects mean that the Internet needs to be supervised and monitored by regulators to verify if the above proposed rules are satisfied. It is a proposition for and from all regulators, even outside the European Union. In addition, it is asked in the EU (where the EC is entitled to impose rules) to forbid operators to block users from accessing legal content. ARCEP has been more stringent, asking for non-discrimination between Internet traffic streams [1].

A last recommendation is that neutrality requirements should be different depending on whether we consider mobile or fixed networks, because (1) capacity is much more limited in the mobile case, (2) because of mobility itself, which has to be handled by operators, and (3) because of the larger type of attached devices (phones, smart-phones, netbooks, broadband cards), complicating the management. Traffic management has, therefore, to be more flexible in the case of mobile networks, due to their specificities [1].

The impact of the decisions made after those consultations is of great importance at the global political level too. Indeed, as highlighted by FCC Commissioner's Robert M. McDowell [5], several countries were waiting for thc USA (and E.U.), and some regulatory decisions could be used to justify and expand their own government authority over the Internet, but with the motivation to limit the freedom of speech. This has to be taken into account when taking decisions.

4 Arguments of Neutrality Proponents: An Idealistic and Humanist View

The Internet architecture is based on the notions of layers, as defined by the Open Systems Interconnection model (OSI model). Almost all layers are managed by independent entities (or providers). The OSI model abstracts the network into seven layers (physical, data link, network, transport, session, presentation, application) and each entity at a given level interacts only with the layer beneath it, and provides "capacities" to the layer above it, the relations between layers being operated thanks to standardized protocols. In the real Internet, this model can be simplified into a model made of four layers [6, 7]: the content layer (web pages, emails, telephone calls, video, etc.), the application layer (the software permitting the content to be developed), the logical layer (TCP/IP, domain names, all the protocols), and the physical layer (cables, routers, servers, wireless base stations). If there is no neutrality, there is a major risk that this successful and cheap model will be destroyed because ISPs may reduce competition by developing exclusivity arrangements with content and application providers, and use proprietary protocols between layers that may harm the development of new applications by third parties. Changing the structure and the way the Internet has been designed can be dangerously harmful. Without neutrality; it is claimed that the last mile ISPs, who have the market power, will discriminate between content providers, favoring the ones that are connected to them, so that end users will have less choice, which is claimed to be economically inefficient (what economists contest, see the next subsection). This market power (whoever is the most powerful) is said to be a reason to introduce regulation to maintain innovation at the content level. Similarly, having an important level of innovation from CPs will foster access demand of end users and therefore increase or maintain revenues of ISPs.

A main principle attached to the networking (disaggregated) layer architecture, and described as a key factor, is that the network can hardly make a differentiation according to the identity. This is related to an important issue developed by the neutrality proponents: the *freedom of speech* on the Internet [7, 8]. If the network is not open, this simplifies a lot its control by governments or political entities willing to limit information diffusion and retrieving the identity of correspondents. The current layered architecture on the other hand allows a relative anonymity, a decentralized distribution (which makes it easy and cheap to implement), the possibility to reach multiple access points, no simple system to identify content, and tools of encryption; this is exactly what is needed by political parties to speak up in an oppressive country. This political aspect is hardly taken into account by neutrality opponents and economists, but we can note that the current infrastructure has very probably helped the recent Arabic revolutions with all the information exchanged. Changing it could harm the development of democratic values worldwide.

In general also (not only talking about freedom of speech), vertical integration of content and access, as aimed by non-neutrality proponents, especially in an exclusive way, is seen as discriminatory by users and by regulators. Considering other communication areas which have a longer history, such as roads, canals, mail, etc., there has always been a requirement of interconnection and to serve the public, even if resources are owned by private companies.

And the argument of ISPs that revenues decrease and investments are not possible anymore are contested because ISPs keep those cost values secret; if they were more transparent, data could be verified and they could be trusted. It is rather seen that ISPs want to earn money from both ends: end users and CPs.

Finally, a last important reason for neutrality is *universality* (as above on international consultations): all consumers are entitled to reach meaningful content, whatever the technical limitations of their service. They are also entitled to attach any type of device to the network; and to run any application. The basic idea behind that principle is to keep all the population informed and able to communicate, and not to increase the gap between rich and poor people. French government for instance is currently trying to set up an access for the poor at a limited price.

5 Arguments of Neutrality Opponents: An Economic View

As presented in the introduction, the initial reason of ISPs to deviate from a neutral behavior is the disincentive to invest in networking capacities because of the reduced revenues, an issue argued to be often forgotten by neutrality defenders who always seem to focus on CPs innovation only.

The most important argument is that *vertical integration*, which consists in bundling content and access into a single service, and which can be exclusive for some type of content, is often shown to be the most efficient economically. It has indeed been demonstrated by the Chicago school of antitrust analysis that vertical arrangements can even be efficient under a monopoly, because a monopolist can gain from a higher total value of the whole platform (made of access plus content). A typical example, in another but related context, is Microsoft that, while in a quasi-monopole on the operating systems of computers, gains from the development of applications because it increases its own value (the equivalent of Microsoft being here either the ISP or the CP). The Chicago school of antitrust analysis has more exactly proved that there are preconditions under which vertical integration might be harmful: 1) the vertically integrated firm must have market power in its market (to be able to use this powerful position), and 2) the market in which the considered provider will be integrated must be concentrated and protected by barriers to entry [7]. It is claimed that none of those preconditions are met in the current telecommunication industry, and therefore vertical integration would be more efficient than keeping the current neutral model. Vertical integration has for example been successful in cable television in the 80s in the USA, with cable operators investing a lot in the network to be able to respond to demand for content.

In addition, the Internet is rather seen now as a media (i.e., content distribution) business, and should then be treated as such rather than as a classical telephony/telecommunication economic business facility. Indeed, users go on the Internet not for broadband in itself, which is rather a mean, but for the content, goods and services that they can find. As a consequence, vertical integration is therefore more relevant here because in such media business, distribution (broadband access for us) is generally bundled with content. We can remark that it is already implemented, even if in a limited way: almost all ISPs provide a portal over which users can access a predefined content: (selected) news, direct access to some content, etc. This does not prevent (yet) the users from accessing the content of their choice, but is an incentive to go to the one preferred by the ISP.

But vertical integration is not a new topic, it has always been in discussion in the Internet industry, and is not a new issue. Notable examples are when Microsoft incorporated its web browser into its operating system, introducing a major debate, or when AOL (an ISP) and Time Warner (a CP) merged.

It is even advocated that, contrary to the intuition, vertical integration and non-neutrality would enforce competition at the network (ISP) level [7]. This is for instance what happened between cable and DSL providers in the USA, who drastically decreased their price because of the competition, at the benefice of end users; they also may fight on content. As an illustration, we can cite the competition between cable and satellite TV in the USA: (satellite) DirecTV managed to get an exclusivity on the Sunday night NFL game, which could then not be accessed by subscribers of cable operators, except if they paid in addition for the satellite operator. This put pressure on the historic cable operators to improve their content, to innovate in this direction and to develop competition. A neutral network would not have permitted the development of DirecTV. In addition, it is anyway unlikely that ISPs would decide to cut the access to major CPs, such as Amazon, Facebook, etc., because they would immediately face protests and pressure from their subscribers.

Similarly, allowing the ISPs to differentiate service would increase the economic efficiency and social welfare, by responding to the heterogeneity of users in terms of preferences. This means applying revenue management techniques [9] to telecommunications as it has been done in the airline industry, hostelling, etc., with a success that has not been discussed in those areas. Users would benefit from this, since they would be able to choose the offer most adapted to their preferences, and it may even drive to more services attuned to segments of the population. It is even claimed that the introduction of regulation could favor some applications against others, an undesirable effect. Also, differentiation and market segmentation would allow many ISPs to survive, since if not implemented, declining costs otherwise usually result in a limited amount of surviving companies and natural monopolies. As a last remark on service differentiation, we can note that the network "naturally" makes some differentiations anyway: a protocol like TCP gives less throughput to users far away, and BGP allows to discriminate at routers.

Opponents of neutrality, even if they admit that the success of the Internet is due to the end-to-end architecture, believe that it is not a reason for not changing it now: it has the right to evolve because of economical changes. Even more, writing in the stone a rule because it initially worked might be really harmful. It is not because the architecture has been optimal that it will continue to be. They rather see this immobility reaction as a scare of change from computer scientists, not ready to lose the "baby" they gave birth to, but the Internet users are in general not highly skilled (in opposition to early subscribers to the network) and are not that sensitive to this issue. As mentioned by A. Thierer [7, Page 94]: "If consumers really wanted a pure dumb Net connection, then why does AOL's walled garden have over 30 million subscribers worldwide while charging $23.90 per month?". We do not know exactly what will be the next business model, but there is no reason to necessarily keep the one in place.

Even at the CP level, vertical integration is said to be preferable because it may help (especially in a segmented market) to push applications that have a limited impact. Using proprietary protocols for development and distribution could simplify the advent of new applications, and thus foster innovation thanks to a larger range of networking tools. And as the above example on the NFL games shows, competitive ISPs have an incentive to help content developers in a non-neutral regime, particularly when there is a monopoly for some type of content. Interoperability, even if appealing *a priori*, is not necessarily efficient from an economic point of view.

Promoting a universal access to the network is also contested by neutrality opponents. They believe that this may harm investment because there is no real incentive in that situation for ISPs to invest in capacity and technologies if access is granted, this even more if service differentiation is not permitted: how to define access prices if access is mandatory? This universal access could maybe be discussed for applications such as web browsing or email, but not for bandwidth-consuming applications. In addition, it is argued that the full connectivity as a principle to maximize social welfare does not take into account an important component: the congestion that end users will experience if there are no sufficient investments on capacities or technologies by ISPs.

This universality principle has been somewhat questioned by governments too. As an illustration, France has passed a law in 2009 that disconnects users from the Internet if they download illegally copyrighted material. Even if this has raised a debate at the law level (the French Constitutional Council asked to modify its implementation because it was in contradiction with the Declaration of the Rights of Man and of the Citizen coming from the French revolution: the ISPs will not decide by themselves to filter traffic, but will implement the decision if it comes from a judicial authority), such a law has been enforced (the so-called HADOPI law), even if the Internet has been seen as a mean to exercise the freedom of speech. This differentiates the treatment of legal and illegal traffic, but the notion of legality being considered differently in other countries, this could go against freedom of speech.

It is also remarked by neutrality opponents that imposing a regulation on neutrality would just add or shift power (even) closer to CPs, especially in the case where there is a limited competition. Typically, Google would become even more powerful with YouTube that already accounts for an important part of the whole Internet traffic.

Even at the security level, a vertically integrated service is advocated to ease the control of attacks (because the current model is more or less based on trust and cooperation between communicating nodes) and limit illegal content, something of interest for regulators.

6 Conclusions

As sketched by our presentation, the network neutrality debate can rather be seen as the opposition between (1) an idealistic (neutral or weakly neutral) network as imagined initially by scientists, with an organization in layers, a low cost, and for which end-to-end connectivity and universality are the key issues, and (2) a purely economic (non-neutral) view of the network. The goals of both worlds are therefore different and make a strict comparison difficult.

A question is then: are end-users and computer scientists just scared of the potential change? Or is it too risky and could it potentially lead to a "telecommunication / connectivity" crisis? Would it be accepted by the public as easily as it accepted the Internet in its current form? We wish the reader to make her own opinion on the subject and hope that our discussion will be helpful towards that goal.

We may mention a particular case where neutral and non-neutral networks coexist: the road network, where you have commercial (highway) roads with tolls, and national roads, with a lower quality of service, but that are free of charge and allow (in most countries) a full connectivity. Those national roads are maintained by states, and one could check how this (economic) model could be transferred to telecommunications.

An important remark to conclude this section is about another neutrality threat, that could be viewed as the most "dangerous" and that is not mentioned enough, the search engine neutrality. To illustrate the problem, Google's CEO, Eric Schmidt, actually had to testify in September 2011 before the US Senate about a possible discrimination in the ranking of the searches on the engine, because it is claimed by some to favor the content provided by Google itself. Typically, but not only, favoring YouTube in front of competitors such as Dailymotion. Search engines are indeed very important actors in the current Internet, being the mean by which most content is reached. Their activity can greatly influence the way the network is used and their behavior has to be carefully reviewed too. Because of that, Google is facing a risk to be dismantled under the antitrust law.

References

1. ARCEP: Neutralité de l'internet et des réseaux: propositions et orientations (2010), http://www.arcep.fr/uploads/tx_gspublication/net-neutralite-orientations-sept2010.pdf
2. Commission of the European Communities: Telecom reform 2009: Commission declaration on net neutrality. Official Journal of the European Communities L337 (2009)
3. Federal Communications Commission: Appropriate framework for broadband access to the Internet over wire-line facilities. FCC 05-151, CC Docket No. 02-33 (September 23, 2005)
4. Federal Communications Commission: Notice of proposed rulemaking. FCC 09-93 (October 2009)
5. Federal Communications Commission: Order in the matter of preserving the open internet, broadband industry practices. FCC (April 2010)
6. Lemley, M., Lessig, L.: The end of end-to-end: Preserving the architecture of the internet in the broadband era. 48 UCLA Law Review 4 (2001)
7. Lenard, T., May, R.E. (eds.): Net Neutrality or Net Neutering: Should Broadband Internet Services be Regulated? Springer (2006)
8. Odlyzko, A.: Network neutrality, search neutrality, and the never-ending conflict between efficiency and fairness in markets. Review of Network Economics 8(1), 40–60 (2009)
9. Talluri, K., van Ryzin, G.: The Theory and Practice of Revenue Management. Springer (2004)
10. Wong, S., Rojas-Mora, J., Altman, E.: Public consultations on Net Neutrality 2010: USA, EU and France. Tech. rep., SSRN (2010), http://papers.ssrn.com/sol3/papers.cfm?abstract_id=1684086
11. Wu, T.: Network neutrality, broadband discrimination. Journal of Telecommunications and High Technology 2, 141–179 (2003)
12. Wu, T.: The broadband debate: A user's guide. Journal of Telecommunications and High Technology 3, 69–96 (2004)
13. Zelnik, P., Toubon, J., Cerutti, G.: Création et internet. Ministère de la Culture et de la Communication, République Française (2010), http://www.culture.gouv.fr/mcc/Espace-Presse/Dossiers-de-presse/

Green Networks and Green Tariffs as Driven by User Service Demand

Louis-Francois Pau

Copenhagen Business School & Rotterdam School of Management
lpau@nypost.dk

Abstract. This section describes an approximate model built from real subsystem performance data, of a public wireless network (3G / LTE) in view of minimum net energy consumption or minimum emissions per time unit and per user. This approach is justified in order to generate the integrated view required for "green" optimizations, while taking into account service demand and operations While subsystems with lower native energy footprints are being migrated into public networks, the many adaptation mechanisms at sub-system, protocol and management levels make system complexity too high to design major comprehensive "green" trade-offs. However, by focusing on the incremental effects of a new network user, the approximate model allows marginal effects to be estimated with good accuracy. This capability allows for the provisioning of personalized energy / emissions reducing tariffs to end users with inherent advantages both to operators, energy suppliers and users. One key advantage is the possibility to reduce waste capacity, and thus energy consumption in the network, by allowing the user to specify just the service capacity and demands he/she has. From an engineering point of view, the incremental model allows to tune sub-system characteristics jointly, especially transceivers, transmission and storage. From a configuration point of view, the model allows to determine which nodes in the network benefit most from back-up and renewable power sources. From a business perspective, the model allows to determine trade-offs between personalized bundle characteristics and the energy cost share in the marginal operating expense share. Detailed sub-system model and design improvements are carried out on a continuous basis in collaboration with industry.

Keywords: Wireless networks, Energy consumption, CO2 emissions, Marginal analysis, Personalized tariffs, Green wireless tariffs.

1 Introduction and Scope

Ever since wireless networks have been developed with mobile terminals, the energy consumption of autonomous access terminals as well as of the access infrastructure has been in focus at the design stage, just to ensure the longest possible communications link durations. When later such wireless networks have been introduced by licensed civilian public operators, the energy expenses of the access and backbone infrastructures have turned into a growing component of their operating expenses,

A.M. Hadjiantonis and B. Stiller (Eds.): Telecommunication Economics, LNCS 7216, pp. 117–125, 2012.

creating the need for thorough energy consumption analysis. This analysis has especially focused on the spatially decentralized layout (and thus power supply) of base stations and transmission links, and on coding/modulation schemes. Later on, operators and equipment suppliers alike have put more and more emphasis on the CO2 emissions from network and device operations, as well as on recycling and sustainable designs ; this last move has so far been motivated by a combination of corporate citizenship, competitive "green" branding, and pressure for better device and system designs. In this evolution, very little attention has been given to the changing nature of the communications traffic carried by the public wireless networks. Likewise the shift in part away from voice calls and messaging, to mobile data with growing effective bandwidth, has not yet impacted wireless network energy analysis. And, most importantly, the user has not been incentivized to drive their traffic and services demands in a way leading to energy and emissions reductions, ultimately making them responsible in the overall processes. This section addresses therefore the following issues:

1. Characterize the energy consumption and emissions jointly of public wireless network infrastructure and of the actual user-selected service derived traffic;
2. Characterize wireless infrastructure technology migration effects on energy and emissions (from UMTS to HSPDA and LTE);
3. Offer a tool to industry allowing to design basic and value-added service tariffs in view of eventual investments in renewable energy production, and of the introduction of "green" wireless telecommunications tariffs.

It has been chosen in this section not to include mobile terminals' and smart phone's own energy consumption and emissions patterns, as the variability thereof is very big due to: technology (incl. operating systems), designs, usage, and battery properties over time. That contribution from terminals to emissions is known to be an order of magnitude less than the one from the networks. The unique service needs enabled by smart phones is addressed through value-added services. Also, in view of the high complexity in network architecture, technologies and operations, approximations have been made sufficient to address issues 1-3 above, which differs from contributions to specific devices, coding schemes, or protocols.

Part 2 addresses the modeling methodology, part 3 the approximate wireless operator energy model, part 4 some examples of numerical results, and part 5 their use for green wireless tariffs. As this work was done for and with industry, there are no public references.

2 Modeling Methodology

It is obviously very difficult with present state-of-the art to build detailed engineering models calculating precisely the overall energy consumption or emissions of a complete wireless network in operations. This difficulty is not unique to the field, in that it is shared with many other large scale systems serving users with mostly services: e.g., transportation, logistics, national economies, and power grids themselves.

However, microeconomic theory comes to the rescue, in that it proposes not to build macro-models for prediction, but marginal models for an individual user assuming an approximate model of the global system, and that the individual user is representative of the aggregate set of users. In other words, the marginal models aim at modeling as accurately as possible, for a given context, the effects of one typical user on the first or second order derivatives of the global model.

This approach can be expanded to consider a distribution of individual users, even with outliers, to estimate the distributions of the marginal effects of a given user, while preserving the fundamental properties of the approximate global model.

In the context of wireless green networks, the following marginal indicators driven by user generated traffic and service requests are especially important:

- the individual user's marginal contribution to total network energy consumption (over a given time frame);
- the individual user's marginal contribution to total CO2 emissions (over a given time frame); and
- the individual user's marginal contribution to network energy costs, network operating costs (OPEX) , and network capital expenditure (CAPEX) in that and traffic request.

The above cost elements can be offset against the incremental revenues dictated by the tariff plan allowing for the requested service and traffic provisioning, so that the network operator's incremental profit (or loss) can be determined. In this section, rather than letting the operator fix the tariff plans, it is the user who specifies his request (generic services, value-added service, contract duration, and maximum price), and let's operators bid for a contract with him/her, according to a reverse auction process. In this way, excess capacity in the wireless network and the corresponding energy consumption can be reduced, in that the subscriber does normally not want to pay for service capacity or offerings he/she do not intend to use. Therefore, the analysis and modeling rests on the following main tasks:

- Build with industry data an approximate *network infrastructure and traffic model*, with energy footprints for all main subsystems (radio, transmission , cooling) and traffic dependent energy consumption (circuit switched and IP); the network allows the provisioning of a set of services (generic and value-added) to a subscriber base;
- Structure the *individual user's contributions* to CAPEX, OPEX, Billing, CRM, Network management, Content acquisition and net energy costs, as a model of the *marginal flows* linked to one additional user, on top of an existing subscriber base;
- Include a *reverse auction bidding process*, whereby the incremental user states the service duration, his/hers basic bundle price offer, and his/hers value-added bundle offer for user-set service duration; operator then must select and configure network resources accordingly allowing *fine-tuning of tariffs and incentives with profitability and emissions constraints.*

3 Approximate Wireless Operator Energy Model

This part describes an approximate model built from real sub-system performance data, of a public wireless network (3G / LTE) in view of minimum net energy consumption or minimum emissions per time unit and per user. The computational implementations exist at different levels: static non-linear, non-linear with average traffic intensities, randomized over users by Monte-Carlo simulation, and determination of energy /emissions levels by value at risk.

3.1 Services and Infrastructure

A detailed and complex realistic approximate model of a wireless service supplier has been designed and built since 2006 with heavy industry involvement, for the limited purpose described in part 2. This model has served in a variety of projects and has shown its flexibility and reasonable accuracy. The services which can be provisioned include:

- Common services: Network management, billing from CDR and IPDR/CRM;
- Generic services: voice, SMS/MMS and metered mobile IP traffic;
- Sample Value-added service: for illustration is used the M-Singing service where a user download songs, and has interactive comments by a paid employee to improve performance;
- Other value-added services studied: Mobile video, technical wireless field support, public ticketing;

Assuming an aggregated subscriber base number, and specified network coverage, the following input-output sub-models have been built on the basis of subsystem data, performance, and energy/cooling consumption supplied by 8 equipment suppliers or operators:

- UTRAN (Radio): RBS and Node-B's for UMTS, EDGE/HSPDA, LTE (100 Mbit);
- Transmission: line cards, Microwave links, ATM over IP, WDM , SONET;
- Backbone: MGW, edge routers, core routers, AAA, signaling;
- Storage: CDR, billing/CRM data, on-demand media, regulated security records;
- Power: electrical grid power supply, local wind power at base station , local solar power at base station, backup local power; and
- Cooling for most large sub-systems.

Network capacity is balanced to meet minimum QoS thresholds (effective bandwidth, blocking probability, content storage capacity) from the aggregate subscriber base assume to have service and traffic request as the individual user. Excess capacity not used by generic wireless services (voice, messaging, mobile Internet) may be used by value-added services; if it is insufficient, extra capacity is provisioned by SLA's at higher usage costs.

Another essential property of the approximate operator model is that it is not technology dependent, in that it is the service and traffic mix request of the user which drives the access technology whose parameters are used. In other words, and with a simplistic illustration: if the individual user only wants a voice service, the access network parameters will be those of a 3G infrastructure supplying him with that service at his required QoS levels , supplemented by a common backbone. Vice-versa, if the individual user only wants a value-added service with high data rates and content access, the access network parameters are likely to be those of an LTE access network. As a result of this property, the reverse auction process drives technology switching in the access network, with associated performances, costs and of course energy consumption.

Obviously, the quality of the approximate model hinges on sub-system data or measurements. Real technical input-output data (power, volume, voltage, frequency, performance) were obtained in most cases, from 8 different worldwide suppliers, for different technologies / generations. When relevant, published statistical regression results (from estimated or usage sub-system data) produced by various research groups or companies were also used. Building premises and eventual separate data center models have been available but not yet incorporated.

3.2 Energy Consumption, Renewables, and Emissions

Each sub-system has normally an energy consumption which is the sum of a static design related part, and of a sub-system usage part, with in addition cooling. As an example of the usage part for radio transceivers, the energy consumption of each is determined to be at a power level driven by both the inter-base station ranges, and by the channel related capacity driven by traffic; further parameters such as frequency range, and spectrum efficiency and modulation types are accounted for if data were obtained.

All core infrastructure, real-time storage and backup transmission links operate on in-sourced electrical grid power. In addition, an exogenous parametric proportion of distributed nodes (RBS, TRX, links) with renewable energy sources (wind, photovoltaic) is assumed. These renewables obviously impact favorably CO_2 emissions as already demonstrated by some operators. A minimum electrical grid or backup supply (batteries or generators) for all distributed nodes is provisioned, in relation to infrastructure nodes' and local traffic power consumption (incl. cooling).

The mix wind/ solar is parametric, but the operator is assumed to own and operate such sources with the full imputation of the corresponding investments to CAPEX. Excess renewable power supply from the wireless network is sold at eventually subsidized rates, reducing operator's total OPEX. Here data on performances and costs were obtained from two Danish suppliers of renewable energy systems.

4 Numerical Results on Individual User Energy Consumption and Emissions Driven by Personalized Services and Tariffs

The numerical calculations required for the calculation of the specified energy and emissions indicators are quite heavy. The further estimation of the probability distributions of these indicators on the basis of distributions of user requests mandates the use of a high performance or grid computing environment. In this part we will only give two cases corresponding to two personalized user requests in a given wireless operator environment, and some parametric curves of some indicators with regards to the user bidding prices for the generic services and the chosen value-added service (which affect the wireless technology GSM, EDGE, UMTS, HSPDA, LTE used to supply the personalized demand).

The operator environment has the following properties: 10 Million subscribers, tele-density of 576 users/ cell, typical average power consumption / user /month of 18 kWh, share of RBS with renewable sources of 15 %.

Case 1: Contract duration of 3 months; user bid for generic services of 50 Euros; user bid for MSinging service of 100 Euros.

With this demand profile, the traffic is mostly carried by EDGE, and the emissions indicators are:

- CO2 emissions driven by capacity in kg/user/contract: 20.02
- CO2 emissions driven by capacity and generic services in kg/ user/ contract: 21.05
- CO2 emissions driven by generic services in kg/user/ contract: 0.58
- CO2 emissions driven by value-added service in kg/ user/ contract: 0.45
- Renewable energy resold by the network during contract duration: 403 581 Euros

Case 2: Contract duration of 1 month; user bid for generic services of 260 Euros; user bid for MSinging service of 50 Euros.

With this demand profile, the traffic is mostly carried out by LTE (100 Mb), and the emissions indicators are:

- CO2 emissions driven by capacity in kg/user/contract: 6,57
- CO2 emissions driven by capacity and generic services in kg/ user/ contract: 7,09
- CO2 emissions driven by generic services in kg/user/contract: 0,52
- CO2 emissions driven by value-added service in kg/ user/ contract: 0
- Renewable energy resold by the network during contract duration: 194 071 Euros

In general the analysis of results proves that here are very many interactions to account for. Although CO2 emissions due to infrastructure capacity/coverage dominate, the share of generic services and especially of value-added services grows rapidly with service/content richness and QoS real-time performances. While taking into account spectral system efficiency and frequency bands, individual emissions get slightly smaller with newer technologies (HSPDA, LTE), subject mostly to design and microelectronics progress; critical is the mix of low emission green technologies in RBS nodes.

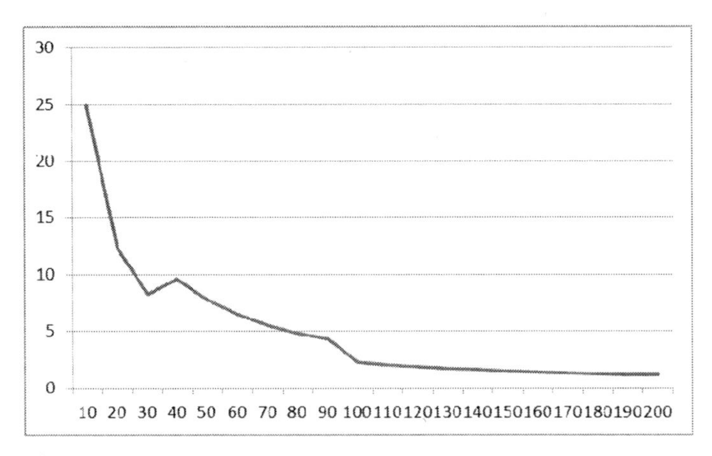

Fig. 1. Net Energy costs per user in % of total OPEX vs. generic services bundle price in Euros; assumes contract duration of 3 months, and a value added service bid of 100 Euros

5 Green Wireless Tariffs

We define here a "green wireless tariff", as the contractual price paid by a subscriber to an operator for a set of services at a given quality of service for a specified duration; this price must reflect both parties best efforts to reduce CO2 emissions, and pass onto the subscriber a share of the energy and emissions savings in the network. The operator can easily offer "green wireless tariffs" by having a green tariff subscriber base, which, against contracting such plans, commits the operator to earmark a specified percentage of their energy consumption at network level to technologies with fewer emissions and possibly to the use of renewable energy supplies.

If furthermore, such green wireless tariffs are made personalized, they provide an incentive to users to specify and adopt a wireless service demand profile leading to less wasted or unutilized capacity, as unutilized capacity burns energy. As this very fact is likely to have a relatively more drastic effect, there is a strong likelihood that personalized green wireless tariffs are to be become very important.

Their deployment however rests on equilibrium between users requesting personalization, and operator's economic sustainability as they then have to make new investments in technology and/or renewable power sources they control. The analysis made possible by the model described in this section shows that operator's profit are not eroded and sufficient to sustain such investments for generic services (Figure 3); however, for value-added services, the situation is less clear as they also demand investments in content, storage, competence (Figure 4). The revenues at network level from the resale to the grid of excess renewable electrical power, obviously increases profits, even if some are shared with subscribers to green wireless tariff plans.

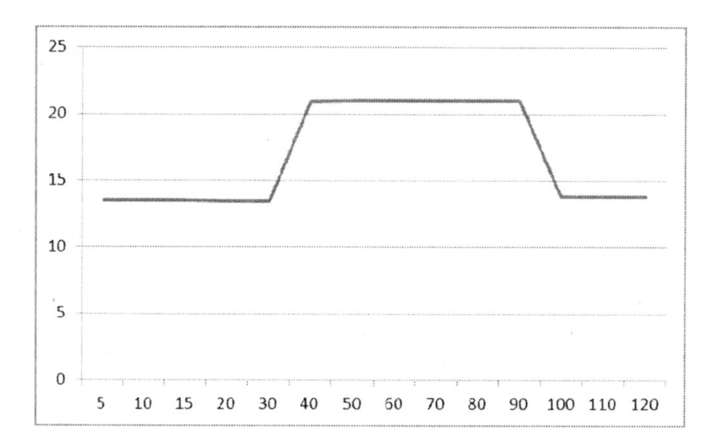

Fig. 2. CO_2 emissions per user in kg CO_2 vs. generic services bid in Euros (access terminals excluded); this assumes contract duration of 3 months, and a value-added service bid of 100 Euros

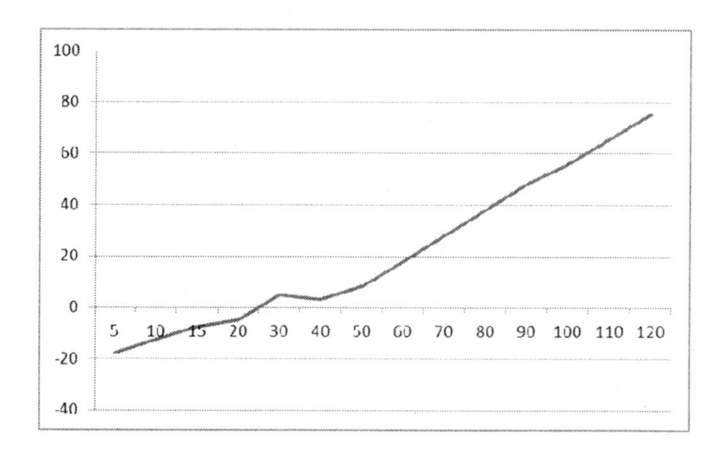

Fig. 3. Net incremental profit per user from personalized green tariff vs. generic service offer in (Euros) for contract duration; contract duration is 3 months and value-added service offer is 100 Euros

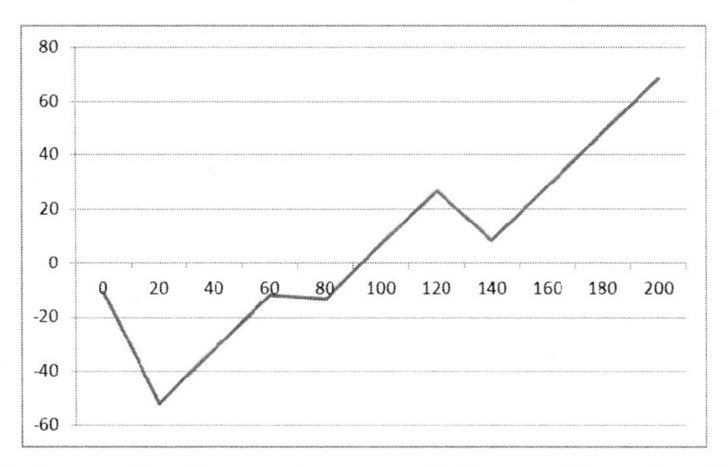

Fig. 4. Net incremental profit per user from personalized green tariff vs. value-added service offer in (Euros) for contract duration; contract duration is 3 months and generic services offer is 50 Euros

In summary, new "green" tariffs can be designed with: (a) personalized service characteristics reducing wasted network capacity, and (b) incentivizing users to larger renewable power supply grades by OPEX bonuses. The impact is though mostly from and for high traffic/ content users.

6 Conclusions and Future Research[1]

By modeling and analyzing wireless networks energy consumption and emissions in order to make explicit and quantify individual user's impact, this section has shown that this integrated modeling approach is necessary as each incremental user has a significant impact. It is also necessary to be able to design and implement "green wireless tariffs" and their personalized versions which reduce waste in network utilization.

[1] Due to the nature of this study no references can be provided.

Environmental Impact of ICT on the Transport Sector

Morten Falch

CMI, Aalborg University Copenhagen, Denmark
falch@cmi.aau.dk

Abstract. This section analyses the environmental impact of ICT. This is done through an analysis on how two specific applications (tele-work and teleshopping) affect transport behavior. These two applications are considered to have a large potential with regard to savings in energy consumption.

Keywords: Green ICT, Tele-work, Teleshopping, Telecommuting, ICT impact on transport.

1 Introduction

The transport sector is both one of the most energy consuming sectors, and also one of the sectors having the largest potential for reduction in negative environmental impacts through use of ICT technologies. In fact environmental concern has been an important factor in driving policies supporting transport related applications of ICT. In particular in the US, ICT has been seen as a way to change a clearly unsustainable trend in the transportation sector without imposing unpopular restrictions on transport.

One of the first studies seeking to calculate the macro impact of ICT on transportation was a study prepared by Arthur D. Little in 1991. The study was very optimistic with regard to the potentials. The executive summary begins as follows: "Can telecommunications help solve America's transportation problems? The answer is definitely yes!"[2]. This answer is based on calculations of the savings, which can be realized through:

- Commuting to work substituted by "telecommuting"
- Shopping, substituted by "teleshopping"
- Business trips, substituted by "teleconferencing"
- Transportation of information, substituted by electronic information transfer.

These calculations estimate that a 33% reduction in transport can be achieved by use of telecommuting. Later studies are far less optimistic regarding the potential savings. One reason is that second order and third order impacts are taken into account (see below). Another is that diffusion of transport reducing ICT applications. Back in the 80's it was expected that as many as 40% of the employees could work as teleworkers [8], but the current level is much below this optimistic figure. Later studies, e.g. [6], have estimated savings at the level of 2-3%. A study of the impact from telecommuting and teleshopping in Denmark estimated the savings to be about 1% of the total person transport [14].

A.M. Hadjiantonis and B. Stiller (Eds.): Telecommunication Economics, LNCS 7216, pp. 126–137, 2012.

This section analyzes the reality of these aspects and reduces some of those uncertainties by introducing visions for different specific applications regarding the three areas; hence it should not to be seen as an analysis of the whole ICT complex. There are four different dimensions of the impact on the environment from the transport sector:

- Overall demand for transport by type
- Demand by transport mode
- Efficiency by mode of transport
- Environmental impact from different modes of transport

While the first two points mainly relate the demand site, points three and four (in particular four) mainly relate to the supply site. First, one must distinguish between freight and person transport as the ICT impact is very different for these two types of transport. For each of these the ICT impact on the overall demand must be assessed. Person transport is usually measured in person kilometers and freight in ton kilometers. ICT can affect the overall demand in various ways: Different applications of ICT may substitute the need for transport for instance by use of tele-work or by use of e-mail instead of surface mail. But ICT may also generate new needs for transport. This impact will most often be realized through indirect effects, for instance through the impact of ICT on globalization. The overall demand for transport includes a wide range of very different transport needs with regard to location, speed, frequency etc. These different needs will often demand use of different modes of transport e.g. train, private car or flight. As the environmental impact from the modes of transport are very different, it is important not to assess the ICT impact on total demand only, but also the ICT impact on the demand for each mode of transport.

ICT is widely used to increase efficiency of transport through more efficient planning. This implies that more ton kilometers (and to a certain extent more person kilometers) can be realized per kilometer driven by a truck, sailed by a ship etcetera. This implies that more transport work can be carried out without increased impact on the environment. Finally ICT can be used for design of more environment friendly means of transport, e.g. less polluting cars and flights. This latter aspect is essentially dealing with design of more environmentally friendly processes of production and is not specifically related to the transport sector. This aspect will not be analyzed further in this section. Here the focus will be on the impact on the overall level of transport realized by each mode. This has also implications on how the impact on the environment is measured. The impact of transport on the environment is a multidimensional parameter including a large number of different environmental factors such as noise, emission of No_x and CO_2. However we will not address each of these parameters separately.

ICT penetrates all parts of our daily life and all parts of the production. It is therefore impossible to assess the environmental impact of every possible application of ICT, which affect our transport behavior. A study on the ICT impact must therefore concentrate on a limited number of parameters. ICTRANS – an EU project on Impacts of ICT on transport and mobility makes a distinction between ICT applications within three different socioeconomic domains (producing, living and working).

Within each of these domains the most important ICT applications with regard to impact on transport are identified (Table 1). Thus, some applications affect transport behavior within more than one socio-economic domain.

Table 1. Mapping application areas into socio-economic spheres [4]

Producing	Living	Working
Logistic Services		
Manufacturing Services		
Customized Services	Customized Services	
Retailing and distribution	Retailing and distribution	
	Teleshopping	Teleshopping
		Tele-working
		Distance Working
		Self-employment

We have chosen to limit our analysis to two different applications:

- Tele-work (including telecommuting, teleconferencing)
- Teleshopping

These applications are more broadly defined than those defined in the ICTRANS study and cover the most important aspects of the ICT impact on transport behavior within the three socio-economic spheres. As far as possible the analysis will include direct first order effects, as well as more indirect second and third order effects which will be presented. The base for this discussion will be a desk research. Furthermore a number of interviews are carried out with different actors in Denmark. The actors are presenting both the areas of research and the developing business.

2 Tele-work

There are many different definitions of tele-work, some definitions take the technology as point of departure and focus on how of ICT is applied in the work process, while others see tele-work as a new way to organize the work (which brings tele-work close to the concept of distance work). Sometimes tele-work includes only working from home and sometimes it includes any work related use of ICT. We will use a rather broad definition, which include all work related activities where ICT facilities are used to facilitate a change in location of work place. This definition of tele-work includes at least six different categories:

- Telecommuting (working from home and thereby avoiding person transport to and from the work place)
- Tele-working centers (working from a tele-center and thereby reducing person transport to and from the work place).
- Teleconferencing

- Mobile tele-working
- Self-employed tele-workers
- Offshore tele-working

In addition to these categories tele-medicine could be added as a separate category, but tele-medicine can also be seen as a special application of teleconferencing although it includes some applications, which are quite different from traditional conferencing. Also e-learning could be added as a separate category (examples of e-learning are provided in Table 2), as learning also may be a working activity. But in this context, it is more convenient to look at e-learning as a sub-category of some of the above categories. There is no reason to distinguish between e-learning from home and telecommuting and e-learning from the work place is, with regard to the impact on transport behavior, very similar to teleconferencing.

Telecommuting is the most classical concept for tele-work. Tele-work includes employees working from home using some sort of telecom facilities to communicate with their work place. Working from home is not a new concept but has taken place for centuries, but use of telecom facilities has made it a much more flexible solution, which can be used for more purposes. Tele-work 96 can be either part time or full time. Looking at statistics on diffusion of tele-work, it is important to take the definition of tele-work into account. It is common to define tele-workers as employees working at least one full day a week from home. But sometimes also employees working from home occasionally or only part of the day are included.

The concept of tele-working centers has in particular been used in US. Employees are allowed to do their work in a tele-working centre providing the similar facilities to those at the central office, but located in a shorter distance to the home of employees. The idea is to reduce commuting into the crowded city centers. Tele-working centers may be located in the suburbs, but tele-work centers have also been established in rural areas. Here the purpose is rather to foster regional development than to reduce transportation.

Teleconferencing includes on-line communication between two or more places of work. Most definitions of teleconferencing demand a video-link between the different locations. This application has so far not been very successful. The reasons for this have been that the technology has been expensive and inflexible. But this will change as broadband connections become more widespread, and it is possible to establish a video-link from the employees own computer. It can be argued that the use of a video-link is irrelevant for a discussion of the impact of transport behavior. But the idea is that a video-link enables types of communication that almost entirely can substitute traditional business meetings.

Mobile tele-workers are employees, which perform most of their duties outside their office. This could be sales people such as insurance agents or employees involved in repair and maintenance or after sale services. Such mobile workers use ICT to support their work and to reduce the need for visiting their work place once or twice a day. Self employed tele-workers are self employed, who maintain a part of their business contacts by use of one or more of the above mentioned concepts for tele-work. This makes it possible to serve customers far away, which in particular is of importance for self employed located in remote rural areas.

Table 2. Examples of e-learning [5], [11]

A master programme in ICT is offered in a co-operation between Aalborg University in Denmark and Ghana Telecom University Colleague in Ghana. The classroom teaching is limited to a few intensive seminars, while the remaining part of the teaching is mediated via the Internet. Even during the seminars e-learning is applied, as videoconferencing is used in most of the lectures. This enables students in Ghana to follow lectures conducted at Aalborg University. Even some of the oral examinations are conducted by use of videoconferencing facilities connecting the two universities. For the past fifteen years, the IEEE has met engineers' need for flexible and affordable materials through videos, CD-ROMs and self-study courses delivered to them by mail. The components of a typical IEEE Self-Study Course include a study guide, textbook, and final exam. These materials are structured to provide clear-cut learning objectives, self-testing opportunities and helpful information. It is expected that web-technology will be applied to provide this type of training in the future [6] provides an on-line example of such a course.

Finally off-shore tele-working should be mentioned, although the direct impact on transport behavior might seem to be less clear cut than for the other categories. Off-shore tele-working includes out-sourcing of certain information intensive service functions to other areas. This could be routine jobs like ticketing and customer handling from call centers, but also more specialized consultancy services may be out-sourced [1]. The relation to tele-work is that these types of out-sourcing necessitate intensive use of ICT for exchange of information and to become economically viable.

The direct impact of tele-work on transport behavior is in both the living and the working spheres. 1) Telecommuting and 2) tele-working centers relate to the living sphere only, while 3) teleconferencing and 6) off-shore tele-work relate only to the working sphere. 4) Mobile tele-working and 5) self-employed tele-workers relate to the transport behavior in both spheres.

2.1 Diffusion of Tele-work

Estimations on the diffusion of tele-work vary considerable depending on the source. Although it must be expected that the numbers of tele-workers are growing, some of the most optimistic estimations dates back to the early 80's. At that time it was foreseen that as much as about 50% of all office workers would be teleworking.[4] The most recent estimates are much more modest as they shows that 13% of the work force in EU15 was engaged in some sort of tele-work in 2002 (compared to only 6% 1999). Out of these 7.4% are home based. The potential seems however to be considerable larger as two thirds are interested in either occasionally or permanent to work from home according to an ECATT survey from 1999 [7]. In the 90s tele-workers were mainly belonging to the higher echelons of the labor market, but following substantial reductions in prices for establishing tele-working facilities more groups are using this opportunity.

Table 3. Types of tele-work (in %). Incidence of tele-work in selected countries and the EU27, 2005 (%) [10]

Countries	% involved in tele-work at least 'a quarter of the time' or more	% involved in tele-work 'almost all of the time'
Czech Republic (CZ)	15.2	9.0
Denmark (DK)	14.4	2.6
Belgium (BE)	13.0	2.2
Latvia (LV)	12.2	1.8
Netherlands (NL)	12.0	1.9
Estonia (EE)	11.8	1.4
Finland (FI)	10.6	1.6
Poland (PL)	10.3	2.3
Norway (NO)	9.7	1.3
Sweden (SE)	9.4	0.4
United Kingdom (UK)	8.1	2.5
Germany (DE)	6.7	1.2
France (FR)	5.7	1.6
EU27	7.0	1.7

It follows from Table 3 the figures for tele-working varies significantly between the countries. It should be noted that tele-working reduces commuting only, when tele-workers work from home full working days. A survey conducted by Danish Technological Institute indicates that there are very few tele-workers working a full working day from home in Denmark [12]. In Denmark a tax incentive for companies investing in home-based PCs to the employees have had a significant impact on the number of tele-workers. The multimedia tax, introduced in 2010, has removed this incentive, so employees must pay 450 Euro per year, if they are provided with any employer paid ICT devices – even if they are used for work related activities only. This has reduced the number of tele-workers dramatically. In Denmark tele-work is usually considered as a basic labor right, while introduction of tele-work in most other European countries is introduced only if it can be justified in financial terms [13], this makes it difficult to make exact estimates on the number of people using tele-work on a regular basis.

2.2 Impact on Transport Behavior in the Sphere of Living

The factors affecting transport behavior can be summarized as within Table 4. Substantial efforts have been made to quantify the transport impact. In particular the substitution effect has been calculated in a large number of studies. Most studies foresee a reduction between 1-3%. According to Mokhtarian, many of the studies tend however to overestimate the impact as they do not include second order and third order effects. She estimates the effect to be less than 1% of the total travel miles [9]. A Danish study from 1996 estimates the potential impact in Denmark to be 0.7% [14].

This figure is confirmed by a more recent unpublished study by Danish Technological Institute (Danish Technological Institute). However, if home working (working from home without use of ICT facilities is included) a much higher impact can be obtained.

Table 4. Environmental impact of tele-work on transport behavior in the sphere of living

First order effects:	Substitution: The level depends on the number of telecommuters, the frequency of telecommuting and the distance between home and work place (or tele-center) Urban sprawl: Reduction in rush hour traffic
Second order effects:	Short term: Impact on non work related transport Impact on transport behavior for other members of the household. Long term: Reduction in number of private cars Changes in habitation More flexible labor market
Third order effects:	Development of public transport Localization of work places Regional development

The short term second order effects (also called the rebound effect) have been included in a study made as part of the EU funded SUSTEL project. This study indicate that a substantial part of the transport savings are nullified by increased transport for other purposes such as shopping and increased transport by other family members. The latter is particular relevant in one car families.

Table 5. Commuting reductions and rebound effect [13][1]

	Denmark	Germany	Italy	Netherlands	UK1	UK2
Reduction in weekly commuting (km)	105	283	242	98	253	61
Addition travel (km)	77	53	33	42	60	15
Rebound effect in %	73	19	14	43	24	25

2.3 Impact on Rush Hour Traffic and Modality

Commuting is characterized by its regularity: It goes to the same destinations at the same time every day. Commuting is the major source for urban sprawl in the rush hours. Increasing use of tele-work from home will provide more flexibility to the

[1] Two case studies were carried out in UK.

commuters so they will be in a better position to avoid rush hours. Tele-work will therefore imply a more equal distribution of the load of person traffic during the day. This will lower problems related to crush during rush hours, but may also add to more traffic in person cars, as this may be the preferred mode of transport outside rush hours. Tele-work may imply that people will accept to travel longer distances or on routes not covered by public transport services once they need to visit their work place. Both will add to less use of public transport services. This is illustrated by the fact that public transport has a market share above average in commuting related transport purposes.

2.4 Impact on Transport Behavior in the Sphere of Working

The factors affecting transport behavior can be summarized as in Table 6. The impact of transport behavior in the working sphere is much less studied than the impact in the private sphere. One reason maybe that with regard to personal transport commuting is seen as the major transport problem - not so much because of its dominant role in the total transport – but rather because it is the major cause to urban sprawl during rush hours. Another reason is that it is much more straightforward to calculate the substitution impact with regard to telecommuting. Business trips are not as regular as commuting trips and will often be longer than commuting trips. It is very difficult to estimate the potential for substitution. In particular if this potential is defined as additional substitution compared to what is done today.

Communication between businesses takes many forms including use of low tech solutions such as letters and phone calls, and a video conference may be a substitute for these types of communications as well as for a business trip. The above mentioned study from Arthur D. Little, is one of the few attempts to estimate the substitution effect. Here it is assumed that 13-23% of all business trips may be substituted. This includes transport related to learning activities.

Table 6. Environmental impact of tele-work ontransport behavior in the sphere of working

First order effects:	The level depends on the number business it is possible to substitute and length of the trips.
Second order effects:	More intensive communication with current business partners. Extension of business networks More outsourcing and specialization
Third order effects:	Out-sourcing and globalization of production. Internationalization of markets Localization of work places Regional development

One example is air maintenance. In this case most of the necessary information is stored in a digitized format, and maintenance and repair decisions can therefore be taken without physical presence of aviation experts. In the same way it may possible to operate robots used for medical operations. E-learning can be implemented by use of e-mail only, but distant conduction of lectures and oral examinations demand more sophisticated ICT applications. The transport implications of mobile tele-working are rather different than those for home based tele-working. Like home based tele-workers mobile tele-workers may avoid commuting to their work place, but they may increase transportation during working hours. A study by BT in which most tele-workers were mobile tele-workers indicates that the total transport work may increase. 18% of the respondents stated that there in work travel increased by an average of 267 miles per week, while 9% stated that it decreased by 394 miles.

2.5 Long Term Perspectives

Tele-work does not require use of advanced ICT technologies, but new technologies may open-up for new applications of tele-work. In the short term mobile technologies will be the most important. 3G will enable more mobile applications of data transfer and video, making it easier to connect not only from home but also from any other location. Security will also be a crucial parameter, as companies still may be hesitant to enable access to sensitive information from outside. In the long term development of high quality video communications offering virtual reality like alternatives to physical presence may be developed to facilitate informal knowledge sharing.

The long term impact will however also depend on how conditions for personal transport evolve. Development of alternatives to physical presence will depend on how difficult it will be to make use of physical presence. If transport is both expensive and time consuming electronic alternatives are more likely to be developed.

3 Teleshopping

The concept of teleshopping covers a wide range of business activities with very different implications for transport and the environment. We will in our discussion focus on three different activities:

- Before sales: Activities related to identification of providers or customers and services/products to be acquired.
- Sales: Processes related to the sales transaction such as payments, delivery, signing of contracts.
- After sales: amongst others customer support and maintenance.

Teleshopping and to a certain extent e-business tend to focus on the sales process itself, but after sales and in particular before sales are equally important functions – also with regard to transport implications. E-business affects both the working and the living spheres. B2B will mainly affect the working sphere, while B2C affects both living and working.

E-business in relation to consumers is in reality the same as teleshopping. Teleshopping is not an entirely new concept. Similar ways of distant shopping has been carried out without use of advanced ICT. Mail order has been used for decades and ordering by phone is also a well-established way of shopping. However, the Internet and a wider penetration of broadband, has enabled a dramatic increase in the potential for distant shopping. The benefits of teleshopping for consumers are:

- More convenient shopping
- Timesaving
- Savings in transport
- Faster delivery
- More transparent markets
- Better market access
- Periphery regions/developing countries
- Physical impaired
- More competition

Table 7. List of impacts on transport behavior from B2C (Adapted from [15])

• Some shopping trips will be replaced by deliveries. • B2C will result in the increase of small-part sending (e.g. courier, express and packet deliveries) to an increased number of end-customers with individual delivery-places and delivery-times • The total number of tours will increase. • B2C-traffic will concentrate on suburban areas. • Storage concepts, distribution and collecting traffic have to be adapted. • Small part sending will require more small vehicles.

More transparent markets are in particular related to before sales processes, where different products and suppliers are compared. This can be done much faster and without any person transport. This may however imply that the consumer become aware of suppliers located long away. If the actual purchase is done in the traditional way, the transport savings achieved through electronic scanning of the market, may therefore be nullified through more transport in the sales process. One of the major barriers towards teleshopping has been establishment of efficient distribution systems. Teleshopping has therefore been particular successful in areas, where the goods either can be transmitted via the telecom network or where they can be delivered via the existing postal mail systems. A study by the German ministry of for Transport lists following effects of teleshopping on traffic (Table 8). The list is made through a survey of a large number of German studies. The list implies that B2C is foreseen to have a wide range of impacts on transportation, both on the overall levels of freight and person transportation, and on the structure of the transport. It is however difficult to derive any firm conclusions on whether transport will increase or be reduced, in particular if second order and third order effects are included.

B2C will increase transport related to delivery of goods and change the structure towards smaller units, but this does not necessarily increase the total needs for transport. According to an American study, best-selling books ecommerce has had a smaller impact on the environment than traditional delivery. Also a Swedish study indicates that e-commerce not necessarily will lead to more traffic. In the long term a wide penetration of tele-shopping may add to the on-going centralization of the retail market, which may imply that both shopping trips and e-shop deliveries will involve longer transport distances.

Table 8. Environmental impact of e-business on transport behavior in the sphere of living

First order effects:	Substitution of transport related to market scanning. Substitution of transport with intangible goods. Substitution of person transport with delivery of goods. Change in freight transport towards smaller units.
Second order effects:	Centralization of shopping facilities Some services only available on-line
Third order effects:	Out-sourcing and globalization of production. Internationalization of markets Localization of work places Regional development

3.1 Quantitative Impact of Teleshopping

The potential transport savings depends of the share of personal transport, which is related to shopping. According to the latest survey of transport behavior in Denmark [3] 22% of person transport is related to shopping. It should however be noted that shopping is often combined with other activities (combined trips). This implies that savings will be smaller than that, even with a 100% substitution. The key in transport savings will be if teleshopping is used for daily necessities, and not only for books, durable consumer goods and other kinds of goods bought with low frequency. So far teleshopping in this area has remained limited in spite of the convenience potential savings in both transport and time.

An estimate of the potential impact of teleshopping in Denmark is as low as 0.3% of the total personal transport [14]. Although the figure is somewhat dated and that the potential may have increased since then (for instance due to a centralization of the shopping structure), it gives an indication of how limited the potential is.

4 Conclusions

This section has analyzed the impact of two ICT applications on the total demand for transport (tele-work and teleshopping). These applications were selected as they were expected to be among the applications with the highest potential for savings in transport. In spite of this it is concluded that the total impact is limited and might be even negative if rebound effects are included in the analysis. Still the application of ICT may be able to offer new possibilities for reducing the negative impact of transport on the environment. ICT is only one out of several parameters affecting transport behavior. A major problem in this context is that ICT both offers possibilities for substitution of certain types of transport, and contributes to ongoing behavioral trends at the individual and societal levels that lead to increases in the overall demand for transport. If the use of ICT is accommodated by other measures aiming at reduced demand for transport, such as heavy taxation, ICT can be used to facilitate a development towards less transport intensive lifestyles.

References

1. Baark, E., Falch, M., Henten, A., Skouby, K.E.: The tradability of consulting services. UNCTAD, New York and Geneva (2002)
2. Boghani, B.A.: Can telecommunications help America's transportation problems? – A multiclient study. Cambridge Massachusetts: A.D. Little (1991)
3. Danish Bureau of Statistics, Transport 2005. Copenhagen Denmark: Danish Bureau of Statistics (2005)
4. ESTO, Impacts of ICTs on transport and mobility (ICTRANS). Seville: The ESTO Research Report (2003)
5. Falch, M., Tadayoni, R.: Practical experiences with using E-learning methodologies at CICT. In: Digital Learning India 2006 Conference, New Delhi, August 23-25 (2006)
6. Havyatt, S.: Impacts of teleworking under the NBN. Canberra, Australia: Access Economics (2010)
7. Hommels, A., Kemp, R., Peters, P., Dunnewijk, T.: State of the art Review—Living. Paper prepared for the ESTO study on the impacts of ICT on transport and mobility. Seville: ESTO (2002)
8. Korte, W., Robinson, S., Steinle, W.: Telework: Present situations and future development of a new form of work organization. North-Holland, Amsterdam (1988)
9. Mokhtarian, P.L., Choo, S., Salomon, I.: Does telecommuting reduce vehicle-miles travelled? An aggregate time series analysis for the US institute of transportation studiesUniversity of California (2002)
10. Parent-Thirion, A., Macías, E.F., Hurley, J., Vermeylen, G.: Fourth european working conditions survey. European Foundation for the Improvement of Living and Working Conditions (2010)
11. Ranky, P.: Introduction to sustainable green engineering, `http://ieee-elearning.org/` (retrieved November 10, 2010)
12. Schmidt, L.B.: The impact of eWork on families and the social consequences. Aarhus, Denmark: Danish Technological Institute (2005)
13. SusTel: Is teleworking sustainable? An analysis of its economic, environmental and social impacts. European Communities (2004)
14. Transportrådet. Distancearbejde og teleindkøb – konsekvenser for transporten distance work and teleshopping - impact on transport behaviour, No. 96.09. Copenhagen: The Danish Transport Council (1996)
15. Zoche, P., B., B., Joisten, M.: Impacts of ICTs on transport and mobility (ICTRANS) annex: The impact of ICT on mobility and transport in the socio-economic sphere of working. In: FhG-ISI (2002)

Cost-Efficient NGN Rollout

Sofie Verbrugge, Jan Van Ooteghem, Koen Casier,
Marlies Van der Wee, and Mathieu Tahon

Ghent University, IBBT,
Dept. of Information Technology (INTEC)
{sofie.verbrugge,jan.vanooteghem,koen.casier,
marlies.vanderwee,mathieu.tahon}@intec.ugent.be

Abstract. We introduce in this section potential business models for the deployment of optic fiber networks. Because Fiber-to-the-Home is deployed at a slower speed than expected, we describe the evolution in several European countries, identify the reasons of the differences, propose a holistic approach to improve the business, and present a game-theoretic model to evaluate the impact of municipality investments on the market.

Keywords: NGN rollout, value network, holistic approach, game-theory.

1 Introduction

The rollout of new fiber access networks or the upgrade towards Fiber-to-the-Home (FTTH) is happening in most Western countries at a much slower rate than expected. The final step in migrating to an all fiber (last mile) access network is one bridge too far for most telecom operators. Nevertheless, FTTH is considered as the solution for access networks in the long run. In order to analyze the different technical and business aspects regarding future proof strategies for the rollout of FTTH, a strong focus must be put on the business and techno-economic evaluation.

This section will focus on three main issues within the techno-economic domain of networks. The first paragraph will describe which business models and value networks can be used to deploy FTTH networks, focusing on the different actors involved (both public and private) and the role each of them plays in the deployment and operations of the network. Several existing European networks will be analyzed and compared, to identify the best practices for the rollout of FTTH. Secondly, we will focus on the deployment itself, and identify a holistic approach that can improve the viability of the business case and lead to an earlier deployment and higher coverage. In the third and final section of this section, the influence of competition on the viability of the business case will be investigated, by applying a game-theoretic approach on a case in which a municipality rolls out FTTH in competition with another network operator upgrading its infrastructure.

A.M. Hadjiantonis and B. Stiller (Eds.): Telecommunication Economics, LNCS 7216, pp. 138–147, 2012.

2 Business Models and Value Networks for a Successful FTTH Deployment

To identify best practices for the rollout of a new FTTH network, we start by setting up a framework to analyze existing networks in Europe, focusing on actors involved and the roles they take up. This framework, the so-called network matrix, will be explained first and later in this section used to identify different business models.

2.1 Network Matrix

The network matrix describes the different roles within the deployment and operations of an FTTH network (Figure 1, based on [9]). The different lifecycle phases of the network (deployment, provisioning of the early subscribers together with deployment, provisioning later on and operations) are found on the x-axis, while the y-axis shows the different parts of the network (backbone, access, building and home). Within each cell of the matrix, several boxes represent the different network layers, which can be easily interpreted as an extension of the classical OSI-model [5] towards fiber-based networks.

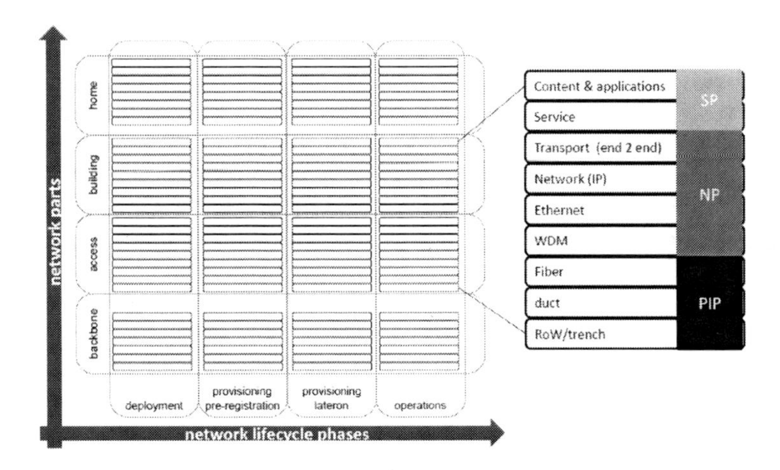

Fig. 1. The Network Matrix indicating all the roles

Focusing on the different network layers within each cell of the matrix allows identifying three important roles, which will be used in the next paragraph to identify the possible business models.

The **Physical Infrastructure Provider (PIP)** is responsible for obtaining Right of Way (RoW, the right to open up the streets) and performing the digging works itself, installing ducts and fibers or blowing the fibers afterwards. The PIP can also take care of other passive infrastructure, such as the housing of the Central Office (CO), installation of empty racks, provisioning of man- and hand-holes and so on.

The **Network Provider (NP)** deploys and operates all the active equipment necessary to provide an end-to-end connectivity between the customers and the CO. They install specific equipment at the CO (Optical Line Terminal - OLT) and at the customer's premises (Optical Network Unit - ONU) and are responsible for the other network equipment (like switches, splitters…) in between.

When an end-to-end connectivity is present, the **Service Provider (SP)** can use the active network to offer services. His responsibility is to install the service-specific equipment (e.g., a set-top box for digital television) and to send the right content and applications to its subscribers. It is important to mention here that we consider only the local service providers that receive direct revenues from their customers through subscriptions and not the over the top players like Google etc. that get their main income out of advertising.

2.2 Business Models

One of the most important issues when analyzing an FTTH deployment is the identification of the business model: who is responsible for which role and does this mapping of actors to roles allow for competition. This paragraph gives a short explanation of the different business models encountered while analyzing the different cases, so they can easily be referred to in the next sections.

Based on theoretical studies and experience from experts in the field, we identified two possibilities for opening up a network to competition: on network level and service level [4]. From these opportunities for competition, several business models can be extracted, but we will only focus here on those that we encountered in analyzing the cases. The first business model is the **Vertically Integrated** one, in which one actor (or its subsidiaries) takes up all the roles. The **Open Access model with competition on service level** is characterized by one single PIP, one NP and several SP's. End-customers are then able to subscribe to services from different service providers, but the provisioning of end-to-end connectivity is a monopoly. The third and last model, **Equal Open Access**, allows for competition on both network and service level, since multiple NP's and multiple SP's are operating on top of one passive network. In this case, price competition can be combined with speed differentiation, because NP's are responsible for the offered bandwidth.

2.3 Existing European Network Deployments

In Europe, there are several networks already deployed and in an operational phase. This paragraph will compare different FTTH networks based on the business model they use, based on the analysis performed in [8]. The cases are subdivided in three types: urban, rural and large-scale deployments, because of the differences in type of rollout and presence of other access networks, like DSL or cable.

Urban Regions

Urban regions are characterized by the presence of other broadband networks, like the DSL network of the incumbent or an alternative network owned by a cable operator.

This might be one of the reasons why we see here a clear preference towards the (Equal) Open Access Business Model, allowing room for competition resulting in a reduction of prices charged.

In urban regions, it is mostly the city (or a publicly owned utility company) that recognized the need for a fiber network and initiated the venture. In general, the main reason for deploying the network comes down to boosting the local economy and the ICT market, as well as increasing competition. Apart from the public entity (city or utility), housing corporations and private investors seem to have a rather big influence, both financially and in aggregation of demand. In Amsterdam for instance, the housing companies took up 1/3rd of the initial investment (1/3rd was taken up by the city of Amsterdam, 1/3rd by private banks), meanwhile ensuring a certain take-up rate from the start, as all their homes got connected. Key motivations for the housing corporations to invest in FTTH networks were two-fold: being able to offer a fast and reliable FTTH connection to their residents, which in turn increases the value of their real-estate property.

Large-Scale Deployments

The second category of European cases under consideration is that of large-scale (country-wide) deployments. Three examples are given: Portugal, Italy and Norway.

In Portugal, it was the incumbent itself deciding to start rolling out FTTH, in order to stay competitive vis-à-vis the cable operator in the digital television market. This case leads to interesting conclusions, as it is often said that the only application that truly needs FTTH networks, is video. Cable operators are ahead because their networks are built to transmit video-services, and they can more cost efficiently upgrade their network bandwidth as well. Clearly this illustrates that there is no such thing as a "killer app", but the use of multiple high-quality video-related services simultaneously is a good motivation to begin to deploy FTTH. Note that the case in Portugal is very similar as what is happening in many Eastern European countries where incumbents are starting with FTTH rollouts due to a lack of good infrastructure.

In Italy, a new company was set up: Fastweb. They saw opportunities in connecting residents in seven municipalities (cities like Milan and Rome and their environments), where the Internet conditions were rather rudimentary. Formulating a partnership with AEM, the electricity company in Milan, they were able to save costs on digging, and in turn gained additional subscribers for the network.

The third large-scale deployment was initiated and fully deployed by Altibox, a subsidiary of the regional Norwegian energy supplier Lyse Energi. This again is a completely different case, with a utility company opting for a "multi-utility" strategy, offering both energy and broadband.

Although the initiators for the three large-scale deployments under study are very diverse, the business model used is the same: a Vertically Integrated model in which one company is responsible for both the passive and the active infrastructure, as well as for the provisioning of the services.

Rural Regions

Rural areas are characterized by rudimentary access to the Internet, and broadband DSL or cable networks are not available everywhere. One could conclude that rural areas form good markets to start deploying FTTH, wouldn't it be that the upfront costs are much higher than in urban areas (because the distances to be bridged in between two homes are much higher). Because of this high upfront investment, and lack of interest from the incumbents to invest, the initiative to deploy the network was always taken by a public institution: a public utility firm or the municipality itself.

Furthermore, it is important to mention that rural areas with no other broadband infrastructure available are the only areas where public funding is allowed (the so-called white areas, as defined in the European Regulatory framework [6]). This again discourages private firms to invest in FTTH in rural regions.

3 Benefits from a Holistic Approach

It is remarkable to see that in almost all cases, it wasn't the incumbent operator that had deployed the network, but a third party. For most European operators, there is more value in upgrading their existing infrastructure than installing a new FTTH network, regardless of the higher operational expenditures. But as soon as one operator deploys FTTH, all other operators will most probably follow in order not to lose their foothold [1].

Therefore, it is important to look for methods to improve the FTTH business case. One approach to improving the FTTH business case could be to use a holistic approach, tackling the business case on three fronts – strategic geo-marketing, synergetic installation and detailed operational modeling, which is the topic of the next paragraph and is based on [2].

3.1 Focus on the Best Customers

The operator needs to know which areas in a region to roll-out first and which areas to postpone or skip. The customer is of vital importance in the outcome of the business case, and the operator will have to select the best set of customers to connect in order to maximize its business case. Working at this level requires a huge amount of information and calculation, and this quickly becomes prohibitive. Building such a marketing strategy requires intelligent clustering approaches aimed at reducing the complexity while not discarding too much detail.

Figure 2 shows the three steps in constructing the best cherry picking strategy for the operator, using both user related information (like demographic, economic and marketing information) and infrastructure related information (e.g., geographic information and information about existing infrastructure). The first step includes gathering all information and classifying all inhabitants according to a limited set of profiles. Secondly, the customers should be grouped in selected areas according to geographical distance and their assumed ARPU as found in their profile. As such we will most probably find different smaller closed areas in the region in which all customers have (or lack) more or less the same drive towards FTTH. Finally, at the end of the first two steps, the operator ends up with a data-set containing for each customer its geographical group and customer profile it belongs to. This information will

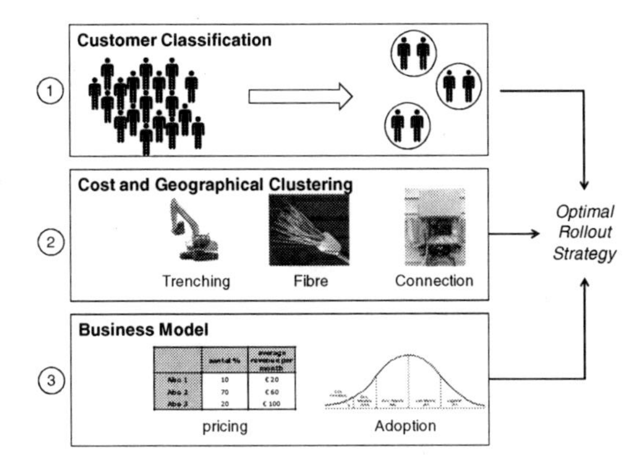

Fig. 2. The three geo-marketing steps for an optimal rollout

form the basis for calculating the cost for the deployment of FTTH in each part of the region. Deploying each group at the right time, taking into account the full business case, will provide the optimal roll-out strategy for the considered region.

3.2 Deploy in Synergy

The cost of installing fibers in the network, also referred to as the outside plant, will dominate (e.g., in a fully buried installation this can amount up to 70% of the overall costs). The business case can be substantially improved by lowering this installation cost. The largest reduction in cost can be achieved by finding synergies with other infrastructure owners for the installation of the network, as shown in Figure 3.

It is worth noting that the synergy can stretch up to the customer connection, in which the customer is connected to all infrastructures in only one intervention. Cooperation between the different infrastructure owners will reduce the final costs charged to the customers as well as reduce the amount of road works in the area.

3.3 Operate the Joint Infrastructure

Installing the outside plant in cooperation with other operators and infrastructure owners will definitely reduce the costs for each actor. On the other hand, such joint installation will undoubtedly lead to important questions considering the operations of the network. Operational expenditures can be up to 50% of the total costs, yet they are often modeled with little or no detail. This uncertain situation poses additional risks.

In order to accurately model the operational costs, two aspects are essential: the flow of activities in the considered processes needs to be detailed and the required input data has to be estimated. It is advisable to use a modeling language (or graphics) that is intuitive to the different people involved in the process (e.g., technicians, experts). Typically flowchart-based approaches are used for this case.

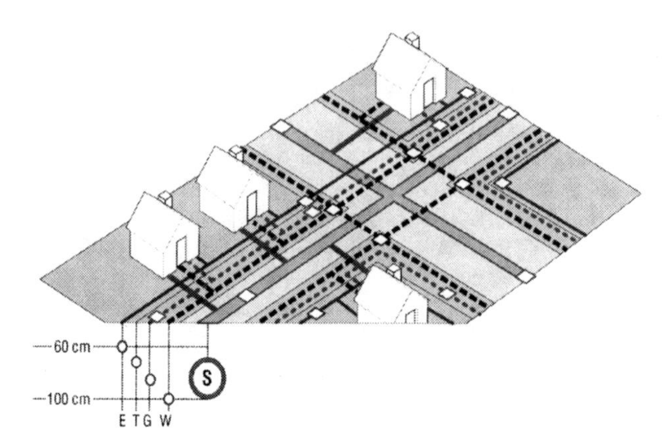

Fig. 3. The short proximity of underground infrastructures (Electricity (E), Telecom (T), Gas (G), Water (W) and Sewage (S))

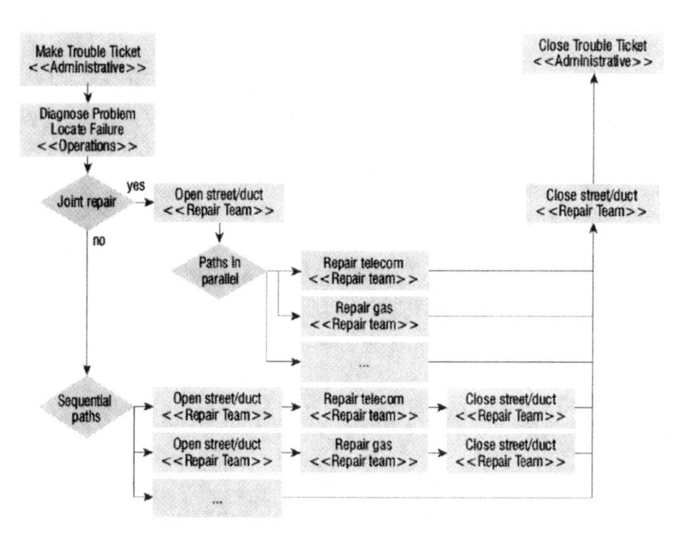

Fig. 4. Flowchart model of the repair process with an indication of the difference between joint and separate repair actions

Figure 4 gives an example for the process model used [1]. This process model shows the actions to be taken when an underground infrastructure is damaged. This model was used as a starting point and a quantitative cost comparison was made between a separate repair process and a joint repair process for all infrastructures - telecommunications, electricity, gas and water. The results showed that, when infrastructures are not close to each other and consequently do not fail together often, this joint process would increase the operational expenditures for the repair. On the other hand, when the infrastructures are close to each other and have a larger possibility for failing together, the joint process can become more cost-effective.

In the situation where the infrastructures always fail together, a cost saving of the repair process (up to 40%) is possible.

It is essential for an operator to model all operational processes in detail. Once modeled, the costs for executing these processes can be estimated quite straightforward. More detailed modeling of all operational processes will also enable the operator to compare trustworthy, different alternative installation types or operations, to find bottlenecks and scheduling problems. As such he can keep a good eye on the current OpEx and control and optimize future OpEx.

4 Game-Theoretic Approach to Improve Results and Conclusions

A third and final issue that will be tackled in this section is the impact of competition with the existing telecom operators on the viability of an FTTH network rollout. Operators do not act in a monopoly environment and must consider the consequences of their actions and decisions on the behavior of their competitors. To model the competitive behavior, game theory is very well suited. Game theory is "aimed at modeling situations in which decision makers have to make specific actions that have mutual, possibly conflicting, consequences" [3]. When it is assumed that all players play at the same time and have sufficiently good knowledge of each other's strategies and payoffs, the game can be represented by means of a payoff matrix. In this matrix, each player has a payoff for all strategic combinations represented in the game. This allows to use tools to find equilibrium states in the game. In an equilibrium state, no player is inclined to change its strategy.

The most commonly known equilibrium state is the Nash Equilibrium (NE), which is defined as the strategy combination in which no player is inclined to unilaterally change its strategy. It is assumed that a game with fully rational players is expected to result in one of the NEs of the game being chosen. Typically static games (the game has one stage in which the players interact) can also be reduced or solved by removing strictly dominated strategies. These dominated strategies have a strictly lower payoff than another (dominant) strategy for all possible counterstrategies. No fully rational player would play a strictly dominated strategy but would instead play the dominant strategy. As such this strategy can be removed for the considered player. By iteratively using this approach for the different players, the matrix of the game can be simplified and can in some case be solved to a unique strategic choice for each player. In these cases, this strictly dominant strategy set is the only NE of the game.

To use the concepts on game theory, it is important to model the impact of the competition between all players on their respective business cases. Most likely, the chosen strategies will have an impact on the division of the market between all players, which in turn will have a significant influence on the potential revenues for each competing operator. In [1], where the strategies consisted of rollout speeds of the different fixed networks, a first mover advantage was introduced to divide the potential market between the cable operator and the FTTH network. Being the first to offer services to consumers resulted in a faster adoption and larger initial market share.

In another case, using game theory to model the competition between different wireless operators, price competition was introduced to divide the market between the different operators [7].

Using game theory and an adequate competition model allows the analysis of the business case of an FTTH deployment under more realistic assumptions, especially since the viability of such a deployment largely depends on the take rate and as such the competition with other operators. If the FTTH network is rolled out in regions where existing communication infrastructures, like DSL and/or cable, are present, this could drastically impact the economic assessment. Applied to the specific case of a municipality rolling out an FTTH network where an existing cable operator upgrades its network to DOCSIS 3.0, game theory offers insight in the most profitable rollout scenarios under competition. Since an FTTH rollout suffers from high upfront installation costs, it is advantageous to start with a slower rollout compared to the cable operator and focus first on the largest industrial sites. In residential areas, FTTH should focus on the densest regions, since this reduces the digging costs per connection. On the other hand, the cable operator has less digging constraints and focuses on the larger residential areas.

5 Conclusions

Broadband penetration is continuously increasing in European countries and operators have to face end users' demand for higher bandwidth. Fiber optical networks are considered as the most future-proof access technologies, but the upgrade towards FTTH is happening much slower than expected. Apart from huge investments needed for fiber deployment other issues improve the viability of the FTTH business case.

First, a framework was proposed to identify different business models: the network matrix. It contains three axes: network lifecycle phases, network segments and network layers, from which the three main roles in an FTTH network deployment can be subtracted: the Physical Infrastructure Provider, the Network Provider and the Service Provider. Several European cases were analyzed and compared, to come to the main conclusion that it most of the existing network deployments, it is a third party taking the initiative for fiber deployment (like the city or a utility network owner). The business models applied vary from being vertically integrated to completely open (for competition).

Although incumbents are currently not investing in FTTH deployments, it is beyond doubt that, if one operator starts deploying the network, others will soon follow in order not to lose too many customers. Therefore, it is important to try to minimize the costs for deployment and maximize the efficiency. This can be done by using a holistic approach, focusing on three steps. First, the operator should subdivide the customers and first connect the most profitable ones. Secondly, synergies with other utility network owners can be agreed upon to reduce the cost for the outside plant deployment. And finally, it is important to not only look at the CapEx, but also model the OpEx in enough detail, so that these costs will not come as a surprise when the network is deployed.

The last subsection of this chapter explained how to investigate the impact of competition on the market share (and thus revenues) that each of the operators can obtain (analysis based on GT). This is especially important to investigate when an FTTH network is deployed in an area where other (copper or cable) networks are already present. Results from a case study showed that when installing a FTTH network in such a region, the deployment should first focus on the densest regions, since this will entail the highest revenue potential.

References

1. Casier, K., et al.: Game-Theoretic Optimization of a Fiber-to-the-Home Municipality Network Rollout. Journal of Optical Communications and Networking 1(1), 30–42 (2009)
2. Casier, K., et al.: Improving the FTTH Business Case. Benefits of a holistic approach. The Journal of the Institute of Telecommunications Professionals 5(1), 46–53 (2011)
3. Felegyhazi, M., Hubaux, J.P.: Game theory in wireless networks: a tutorial. EPFL Technical report, LCA-REPORT 2006-002 (2006)
4. Forzati, M., Larsen, C.P., Mattsson, C.: Open access networks, the Swedish experience. In: International Conference on Transparent Optical Networks (ICTON), Munich, Germany (2010)
5. Kurose, J.F., Ross, K.W.: Computer Networking: A top-down approach featuring the Internet, 4th edn. Addison Wesley (2008)
6. Queck, R., et al.: The EU Regulatory Framework Applicable to Electronic Communications. Telecommunications, Broadcasting and the Internet: EU Competition Law and Regulation (3rd Revised edition). S.a. Maxwell, London (2010)
7. Tahon, M., et al.: Municipal support of wireless access network rollout: a Game Theoretic Approach. Telecommunications Policy 35(9-10), 883–894 (2011)
8. Van der Wee, M., et al.: Measuring the success rate of fiber-based access networks. Evaluation of the Stokab case and comparison to other Western European cases. Institute of Telecommunications Professionals 5(4), 22–31 (2011)
9. Verbrugge, S., et al.: Research Approach towards the Profitability of Future FTTH Business Model. Future Network and Mobile Summit, Warsaw, Poland (2011)

Software Business in the Telecommunications Sector

Lauri Frank, Eetu Luoma, Oleksiy Mazhelis, Mirja Pulkkinen, and Pasi Tyrväinen

Software Industry Research Team, Department of Computer Science and Information Systems, University of Jyväskylä, Finland
firstname.surname@jyu.fi

Abstract. Operations and Business Support Systems (OSS/BSS) software of Communication Service Providers (CSP's) can be developed internally within the CSP or acquired from a Software Vendor. The software industry lifecycle model hypothesizes that software development is internal in the beginning of the industry's lifecycle, and that the share of external products increases when it matures. Empirical evidence shows signs of the OSS/BSS software industry approaching maturity. Current and future developments of the industry include the possibilities of utilizing the Software-as-a-Service (SaaS) model and Open Source software (OSSw). Both have gained increasing interest by the CSP's. However, the relatively small markets (number of CSP's), added with complex and proprietary interfaces of technology and software, seem to hinder this development. The results of scenarios on the future of the industry propose that the adoption of SaaS and OSSw could be motivated by general cost-cutting, by alliances within the industry or by a focus of CSP's business on either providing network capacity or value-added services.

Keywords: Software Business, Industry Evolution, Software-as-a-Service, Open Source Software, Communications Service Providers.

1 Introduction: Software Markets and Their Evolution

In the context of vertical software industry, software business takes place in a dyadic relationship between a vertical industry enterprise and a vendor providing software products or services [6][14]. The vertical industry enterprise usually has its own unit or employees to produce certain software-related functions. Alternatively, the enterprise may find it more efficient and flexible to outsource software development, deployment and operating to an external vendor. Unique applications are more likely to be produced internally, whereas common applications based on established technology are rather outsourced. A common development in vertical software industries is that once unique and differentiating software depreciate into a commodity [15].

The evolution can also be explained through concepts of vertical integration and horizontalization. According to the industry lifecycle model [14], a software market is vertically integrated in the beginning, but eventually disintegrates into horizontal layers of software vendors, whose software products are interacting over standardized

A.M. Hadjiantonis and B. Stiller (Eds.): Telecommunication Economics, LNCS 7216, pp. 148–155, 2012.

interfaces. There are, however, multiple factors that may hinder the above process of horizontalization. These include, among others, technological immaturity and the need to span innovation coordination across multiple software modules and layers, the need for customer-specific tailoring of the software, and the need to maintain compatibility with legacy systems.

The developments predicted by the commoditization viewpoint and the industry lifecycle model are observable in the domain of telecommunication and in the software market serving Operations and Business Support Systems (OSS/BSS) for Communication Service Providers (CSP). A majority of CSPs, who have traditionally themselves developed and maintained all the software needed to run their operations, have recently adjusted their business strategy to concentrate on their key competence areas, i.e. communications and customer intimacy. Thereby they have also outsourced software development and maintenance to the software vendors and system integrators providing OSS/BSS products and services [3].

Furthermore, empirical data suggests that this OSS/BSS software market is reaching a maturity phase: There are signs of market consolidation [3] and software applications like CRM and billing have commoditized to the level that they can be provided and deployed utilizing a Software-as-a-Service (SaaS) model [6]. On the other hand, the results also indicate that the market segments of the OSS/BSS software market might be in different evolutionary stages [3][6] and certain other software applications are developed customer-specific and deployed on-premise. Evidently, the factors hindering horizontalization are present at the software market.

In the following, the attention is focused to more recent and future developments of the OSS/BSS software market. We consider the Open source software and Software-as-a-Service phenomena, which are seen as ramifications of commoditization and horizontalization of the software market. First, we present findings on adoption of software-as-a-service and open-source software in the telecommunication domain. Then, we discuss obstacles of horizontalization. We conclude a presentation of trends and predicted future developments of the software business within this vertical industry of telecommunications.

2 Adoption of Software-as-a-Service and Open Source Software

Software-as-a-Service (SaaS) is a type of software-based service where a service provider offers access to highly standardized commodity software over the Internet to several end-users [6][2]. This is accomplished by running a single instance of the particular software on top of multi-tenant infrastructure [6][11]. The definition implies that SaaS in its pure form is a model characterized by high volumes, high scalability on both the technology and the business model and on-demand pricing [13].

Contemporary literature on SaaS is mostly conceptual, suggesting pros and cons of the model. The economic benefits, less need for internal IT resources, and lower initial and total costs, are associated with the deployment and distribution model of SaaS, enabling service provider to achieve economies of scale [5]. In addition, customers may gain flexibility advantages such as prompt deployment, scalability, easily accessible updates and patches and, additionally, strategic benefits like increased bargaining power over vendors [11][16].

Open source software (OSSw) generally refers to software, which source code is made available to all users along with the usual run-time code and which has licenses with relaxed or very few restrictions. The recent literature has observed the adoption of OSSw in organizations, suggested benefits and problems of OSSw adoption and examined which kind of OSSw are deployed and through studying the antecedents of adoption [16][12][1][4].

Literature suggests that the advantages of OSSw are associated with reuse: customers gain value from reduced time to market, reduced R&D costs, improved process predictability and increased product quality [1]. Comparably, the decision making in end-user organizations culminates on assessing potential cost benefits, on opportunities to exploit communities' resources and knowledge, and on functionalities and maturity of the software under consideration [4].

Using empirical data, we studied the adoption of OSSw and SaaS models in the telecommunication industry [6]. The study applied case research approach including a total of eight companies, six communication service providers and two software companies serving the vertical software industry. We wanted to find case companies of different sizes and breadth of operations and finding markets within telecommunication industry with different phases of maturity. Consequently, European and Chinese communication service providers were selected as the target group of this study. The case study was executed in the year 2010, using two main sources of information: public documents and interviews.

In the examined domain, SaaS adoption is mostly connected to the cost benefits and principally to the flexibility of SaaS offering. Many interviewed service providers suggested that the ease of procurement, ease of maintenance and fast time-to-market are the most important benefits of this mode of deployment. The Chinese service providers are not currently applying SaaS, but the software systems are acquired as custom deployments. This was explained by the interviewed software vendors; software vendors currently have a strong customer lock-in and, thus, low motivation to offer SaaS [6].

Common worries related to SaaS included integration and security issues [6]. For instance, CSPs are obliged by law to apply high data security measures on call data records, which SaaS vendors are not able to comply with. Problems with integration are related to the properties of SaaS offering. The mode of deployment assumes standard processes and interfaces, which does not match the attributes of industry-specific software.

Despite the problems, the SaaS mode of deployment has already been adopted in several companies in the telecommunication industry. Deployed software are horizontal, e.g. for financial management and customer relationship management. SaaS is also in use in the business support systems. However, in the companies interviewed, SaaS is not applied for industry-specific software [6].

With regards to third-party software, both the European and Chinese CSPs see their role in the value chain as reseller and operator of the services, and have already taken such role. One strategy in providing third-party software is to increase customer lock-in, by providing a combination of IT and communication services, and envisions operating in both intermediating and aggregating roles, having already launched product concept to do so. Another strategy is aiming for an aggregator role, where CSP offers multiple SaaS products for end-users. Such role is seen as a natural one, and CSPs are expected to take such role in its ecosystem [6].

The perception and attitude towards open source software is getting more positive in the domain, and the OSSw alternative is analyzed against similar criteria as proprietary alternatives when acquiring software systems. Nevertheless, the current adoption of OSSw among interviewed CSPs could be described as moderate. The OSSw alternatives are used mainly in infrastructure software, including operating systems, application servers and databases. We found that only a few applications specific to this industry exist, namely in service monitoring tools and in voice communication servers built on IP networks [6].

Also, the CSPs are cautious in using OSSw components in systems visible to masses of subscribers and in systems that are otherwise critical to the business, i.e. in systems that require carrier-grade quality and performance. In such systems, the proprietary alternative is often preferred. The main reasons for using proprietary software systems were the lack of capabilities and the required support services. Especially the European CSPs informed that they had outsourced most of their software-related activities. This has led to the organizations not being able to maintain internal capabilities on OSSw. If CSPs deploy OSSw, they will need to contract comprehensive support services from the vendor. As a result, a major share of the cost advantage of OSSw is lost [6].

3 Obstacles to Horizontalization in Telecommunications Software

The need to tightly control the innovation spanning multiple software modules is one of the factors hindering horizontalization. Whenever technology is changing frequently (consider e.g. network technologies advancing from GSM towards LTE and incurred changes in the OSS/BSS software), or whenever the network of inter-dependencies in the system is complex, improvements at the system level require coordinated efforts on changing and refining individual modules in the system. These efforts are easier to coordinate when the modules are under the development in and control of a single company. As a result, the software vendors are reluctant to outsource interconnected modules, and therefore the software development keeps largely concentrated in individual vertically integrated firms, thereby delaying the process of OSS/BSS software market horizontalization [10].

The OSS/BSS software market has also a restricted number of customers, limited by the global base of CSPs. The OSS/BSS software systems are relatively complex and usually need to implement a number of interfaces – both the proprietary protocols of specific network equipment vendors and standardized protocols. Due to this, the costs of developing and maintaining such systems are relatively large, too. These costs are even higher for the software vendors who develop the software products, due to the extra expenses of making the software unified and reusable among different customer CSPs. Furthermore, the software vendors need to add a margin on top of its costs. In the OSS/BSS software market, characterized by high complexity and large number of interfaces to integrate, the revenues from the restricted number of customer CSPs may not be sufficient to recover the large costs of development and maintenance incurred by the software vendors. As a result, the process of vertical disintegration in these segments is unlikely to happen [8].

The complexity of OSS/BSS software systems, and the number and heterogeneity of interfaces to conform with, have also an implication for the type of the software vendors that are capable of providing such systems. For instance, for OSS/BSS mediation software systems, the efforts for implementing interfaces used by an incumbent CSP are an order of magnitude greater as compared with the efforts of implementing the interfaces for commencing CSPs who may require only a standardized subset of interfaces to be implemented. The high efforts required for serving incumbent CSPs are likely to create a barrier for the small software vendors entering the OSS/BSS market. Therefore, the sub-market of incumbent CSPs is likely to be served by large software vendors, who, due to the restricted size of the sub-market and high development costs are likely to remain vertically integrated. On the other hand, the small software vendors will likely serve the sub-market of commencing CSPs where, due to the presence of few standard interfaces and lover costs, vertical disintegration is more likely to take place [9].

4 Trends of Software Business in the Telecommunications Sector

The communications industry has been consolidating mainly through mergers of established CSPs. Simultaneously, new competitors from IT and software industries have penetrated the market, along with Virtual Network Operators (VNO) competing with innovative business models. As a consequence of declining revenues from the traditional business especially in the mature markets, the CSPs' strategic focus is partly on operational efficiency and partly on new sources of revenue through compelling new services. On the other hand, traditional Mobile Network Operators (MNO) are still committed to building and developing their basic network infrastructure, as mobile Internet is increasing the demand for capacity [7].

In line with attempt to achieve efficiency, a clear trend has been that CSPs outsource development, deployment and operating of software systems. The network element manufactures have a strong position in selling software closer to the network interfaces. However, the market for these operations support systems is slowly declining and commoditizing. For business support systems, like Customer Relationship Management (CRM) and billing, new software vendors have appeared to compete with the existing ones. Thereby, analogously to the host industry, the OSS/BSS software market is affected by price competition leading to cost pressures. As a result, the vendors are striving to keep their positions and searching for new ventures in service-based businesses. To strengthen their positions, network element manufacturers and software vendors are engaging in long-term managed service contracts with the CSPs. New sources of revenues are being sought from cloud computing and related services [7].

Both in mature and developing parts of the telecommunications industry, outsourcing of activities related to software systems has increased over time. By outsourcing software activities, the CSPs intended to lower costs and software vendors obviously must produce the services at an even lower cost. According to a CSP interviewee, these cost pressures influence the quality of the software negatively. The CSP may thus lose in competitiveness due to substandard software and services, whether it shows as more limited functionality, a less faultless implementation or poorer performance. In the future, CSPs may therefore make a more prudent evaluation on which activities are outsourced and rather in-source those that they see critical for keeping customers and creating new competitive benefits [7].

Moreover, some of the CSP respondents stressed the skills and capabilities related to different network technologies, while others emphasized understanding customer needs and customizing technologies to match the present needs. Adding capabilities for providing new services was also mentioned. It is consequently natural to observe the CSPs' attempts to create competences through both product innovations and business innovations: In the future certain CSPs may build their competences by developing network assets, technologies and internal resources. These companies may seem inactive, but may maintain their positions by offering network capacity efficiently. Also, certain CSPs may instead choose to compete with contemporary and diligently segmented services built on existing infrastructure [7].

5 Conclusions: Future Scenarios on Software Business in Telecommunications

The adoption of SaaS in only certain types of software, and non-adoption in certain others, indicates that there are factors in the operating environment and in the software business setting, which simultaneously drive and inhibit adoption of SaaS model. The case research [6] revealed that the decision-making on software procurement in communication service provider firms is presently business-driven. There are concurrent pressures to reduce expenditures on software and to deliver compelling services of highest quality. Such pressures drive both adoption of OSSw and SaaS offering and acting as sales channel for third-party SaaS offering. On the other hand, it can be stated that OSSw mode of development and SaaS mode of deployment is not harnessed in industry-specific software, i.e. operations support systems. Reasons mentioned by the CSPs for using the existing systems included specificity of processes and technology interfaces. The standardization and proprietary nature of interfaces is likely to hinder the horizontalization and affect the market structure of software vendors in the OSS/BSS software markets [10] [8] [9].

The trends and developments in the telecommunications and software industries are likely to create pressures towards increasing outsourcing of software-related activities whereas simultaneously CSPs consider in-sourcing certain functions they deem important. Likewise – both product and business innovation – shall be important to achieve and sustain a competitive advantage. Therefore, as alternative or also coexisting future developments one can consider the following scenarios [7]:

1. Cost-cutting scenario: This scenario assumes a continuous increase of outsourcing. This baseline scenario occurs as a result of consolidation in both the host industry and the communications software market. Consolidation enables economies of scale, and both CSPs and software vendors are focusing on lowering operational expenditures. In this scenario, the main driver for OSSw and SaaS adoption is cost efficiency.
2. Alliances scenario: In contrast to the first scenario, operators may become more active by increasing internal software development and related capability in order to benefit from the available standard solutions, OSSw and SaaS mode of deployment. One of the software vendors will attempt to gather its customers into an alliance, which will share software development resources and related costs. In such setting, parties may also share software code and use a standard application developed in the alliance and deployed as SaaS.
3. Bundles scenario: In this scenario the service providers will evolve into two basic types: one focusing on providing network capacity ('bit pipe operators') and the other exploiting their brand to sell value-added services on top of their existing infrastructure. The CSPs profiling them with value-added services will organize their service delivery platforms in a way that third parties (e.g. software vendors) may utilize the platform in a flexible and agile manner. Cloud computing technologies will be in a central role in providing new services efficiently. In cloud computing, mature and widely adopted OSSw solutions already exist. Present considerable projects include e.g. Eucalyptus, a software platform for the implementation of private cloud computing, and Hadoop enabling the creation of data-intensive distributed applications.

References

1. Ajila, S., Wu, D.: Empirical study of the effects of open source adoption on software development economics. Journal of Systems and Software 80, 1517–1529 (2007)
2. Armbrust, M., Fox, A., Griffith, R., Joseph, A., Katz, R., Konswinski, A., Lee, G., Patterson, D., Rabkin, A., Stoica, I., Zaharia, M.: A View of Cloud Computing. Communications of the ACM 53, 50–58 (2010)
3. Frank, L., Luoma, E.: Telecommunications Software Industry Evolution. Presented at the Conference of Telecommunication, Media and Internet Techno-Economics (CTTE), Paris, France (2008)

4. Glynn, E., Fitzgerald, B., Exton, C.: Commercial Adoption of Open Source Software: An Empirical Study. In: International Symposium on Empirical Software Engineering (2005)
5. Jacobs, D.: Enterprise Software as Service: Online Services are Changing the Nature of Software. ACM Queue, 36–42 (July/August 2005)
6. Luoma, E., Helander, N., Frank, L.: Adoption of Open Source Software and Software-as-a-Service models in the Telecommunication Industry. Presented at the 2nd International Conference on Software Business (ICSOB), Brussels, Belgium (2011)
7. Luoma, E., Riepula, M., Frank, L.: Scenarios on Adoption of Open Source Software in the Communications Software Industry. Presented at the 2nd International Conference on Software Business (ICSOB), Brussels, Belgium (2011)
8. Mazhelis, O., Tyrväinen, P., Frank, L., Viitala, E.: Modeling the Impact of Software Development and Integration Costs on the Software Market Size. Presented at the WWW/Internet Conference, Freiburg im Breisgau, Germany (2008)
9. Mazhelis, O., Tyrväinen, P., Matilainen, J.: Analyzing Impact of Interface Implementation Efforts on the Structure of a Software Market: OSS/BSS Market Polarization Scenario. In: 3rd International Conference on Software and Data Technologies (ICSOFT 2008), Porto, Portugal (2008)
10. Mazhelis, O., Tyrväinen, P., Viitala, E.: Analysing Software Integration Scenarios: the Case of Telecommunications Operations Software. International Journal of Management Science and Engineering Management 3, 200–210 (2008)
11. Mell, P., Grance, T.: NIST Working Definition of Cloud Computing. National Institute of Standards and Technology, Information Technology Laboratory (2009)
12. Nagy, D., Yassin, A., Bhattacherjee, A.: Organizational adoption of open source software: barriers and remedies. Communications of the ACM 53, 148–151 (2010)
13. Tyrväinen, P., Selin, J.: How to Sell SaaS: A Model for Main Factors of Marketing and Selling Software-as-a-Service. Presented at the 2nd International Conference on Software Business (ICSOB), Brussels, Belgium (2011)
14. Tyrväinen, P.: Model for Evolution of a Vertical Software Industry. In: Tyrväinen, P., Mazhelis, O. (eds.) Vertical Software Industry Evolution - Analysis of Telecom Operator Software, pp. 25–33. Springer (2009)
15. van der Linden, F., Lundell, B., Marttiin, P.: Commodification of Industrial Software: A Case for Open Source. IEEE Software 26, 77–83 (2009)
16. Ågerfalk, P., Deverell, A., Fitzgerald, B., Morgan, L.: Assessing the Role of Open Source Software in the European Secondary Software Sector: A Voice from Industry. Presented at the First International Conference on Open Source Systems (2005)

Introduction to Future Networks Management Architectures and Mechanisms

Iztok Starc

Faculty of Computer and Information Science, University of Ljubljana, Slovenia
iztok.starc@fri.uni-lj.si

In this chapter, current issues of the Internet are addressed and arising problems of future Internet from socio-technical point of view are tackled. The future Internet considered as a Next Generation Network (NGN) will provide integrated services, i.e. interactive, multimedia, data, real-time, and mobile services. The focus will be placed on the end-user for whom this information and communications technologies (ICT) are primarily designed for. The user's dependence on ICT is rising, being tied to increasing complexity of critical infrastructures (CIs) interdependencies. No wonder that also other stakeholders (service providers and network operators) have to be in the focus.

The future Internet as NGN will only be successful if requirements of users, service providers and network operators are taken into account, and economic and legislative aspects are considered as well. In addition to providing equality of access right so that end-users have the opportunity to choose the most effective service provider, marginalized groups have to be considered as well. They are entitled to participate in social life and thus, broadband has to be provided not only in densely populated areas but also in rural areas. According to aforesaid requirements, we propose several socio-technical solutions, possible alternative choices or improvements to existing ones that will be presented in the chapter outline below, all intended to improve productivity and efficiency of services, security of stakeholders and end-user satisfaction. These selected issues will involve one or more stakeholders and will be related to ICT problems.

In Section 1 on "Economics of Quality of Experience" by P. Reichl, B. Tuffin, and P. Maillé, focus is on the human factor, i.e. quality of experience and its economic implications. This section claims that QoE is a better representation of user's perception than traditional Quality of Service estimation. A better understanding of links between the quality as delivered by the network, as perceived by the user, and as valued by the market can be justified by the logarithmic utility.

Electronic business in the Internet has become an important driver for economic growth. The work on a "Decision Support in Contract Formation for Commercial Electronic Services with International Connection" in Section 2 by M. Waldburger and B. Stiller addresses jurisdiction and applicable law issues of international contracts between consumers and service providers in the case of electronic services. This section proposes a decision support system which provides recommendations on contract parameters at the time of contract formation.

A.M. Hadjiantonis and B. Stiller (Eds.): Telecommunication Economics, LNCS 7216, pp. 156–157, 2012.

Section 3 on "Competition among Telecommunication Providers" by P. Maillé, P. Reichl, and B. Tuffin analyzes a competition between providers implementing a congestion-pricing scheme in various real-life scenarios. The convergence of networks, where Internet, wired and wireless telephony and television are regrouped into a single network, poses additional economic challenges related to providers' competition. In this situation, each network operator has to adapt its pricing scheme in order to attract customers, to maximize revenue and/or to allow fairness in the way resources are shared, but taking care of the decision of competitors who can lure customers.

Section 4 by F. Hecht and B. Stiller on "Economic Traffic Management: Mechanisms and Applications" proposes economic traffic management mechanisms (ETM) to enable competitive overlay services and their management that in order to achieve benefits to end-users, service providers and network operators applied to live and on-demand video streaming as a promising use case. Due to large traffic increases, service providers and network operators need to control and manage network traffic stemming from overlay-based applications.

Section 5 on "Autonomic Management of Mobile and Wireless Networks" by A. M. Hadjiantonis summarizes cooperating autonomic management mechanism for wireless and mobile networking in the future Internet. The principles of autonomic management and self-organization are envisioned by researchers as a natural path for the Future Internet. Autonomic management (or self-management) capabilities aim at relieving service providers, network operators and end-users from tedious configuration and troubleshooting procedures.

Section 6 on "On the Provision of Advanced Telecommunication Services in Rural Areas" by L. Lambrinos focuses on telecommunications in rural areas. Telecommunication operators are not prepared to invest in areas that are not profitable and have only a small number of inhabitants, or places that are remote and infrastructure development costs are high. As a result there are many areas in the world lacking access to the latest telephony and internet technologies. For two different real-life scenarios, a solution is proposed for providing marginalized groups access to features that would not have been available otherwise.

The work on "Electrical Power Systems Protection and Interdependencies with ICT" in Section 7 by G.M. Milis, E.Kyriakides and A.M. Hadjiantonis introduces the need to protect critical infrastructures in order to maintain the economic and secure wellbeing of our society. In this context, the interdependencies between ICT and Electric Power Systems are examined.

Section 8 on "Towards Quantitative Risk Management for Next Generation Network" by I. Starc and D. Trček addresses the reasons why information systems security situation is worsening at an alarming rate and how can proactive, quantitative and computerized risk management approaches overcome this problem, and make information security measurable and manageable.

Economics of Quality of Experience

Peter Reichl[1], Bruno Tuffin[2], and Patrick Maillé[3]

[1] FTW Telecommunications Research Center Vienna, Austria
[2] INRIA Rennes Bretagne-Atlantique, Campus de Beaulieu, France
[3] Télécom Bretagne, Cesson-Sévigné Cedex, France
reichl@ftw.at, bruno.tuffin@inria.fr,
patrick.maille@telecom-bretagne.eu

Abstract. While the recent strong increase of interest in Quality of Experience both in industry and academia has managed to place the end user again into the center of service quality evaluation, corresponding economic implications have not received similar attention within the research community yet. Therefore, in this section we point out some of the key links between the quality as delivered by the network, as perceived by the user, and as valued by the market. The example of logarithmic utility functions allows demonstrating how this broad interdisciplinary approach is able to provide significant contributions in describing and analyzing future telecommunication ecosystems in a holistic way.

Keywords: telecommunication ecosystem, utility function, Weber-Fechner Law, WQL hypothesis, charging for QoE.

1 From Quality of Service to Quality of Experience

While the notion of "Quality of Experience" (QoE) has managed to become one of the key buzzwords in the networking community over the last few years, it may be surprising to learn that the concept itself is far from being new. Indeed, already back in 1994, ITU-T Rec. E.800 [5] has defined service quality as "*the collective effect of service performance which determines the degree of satisfaction of a user of the service*". However, research on "Quality of Service" (QoS), as it has been called since, started to neglect this explicit user-centric focus very soon and instead has put a clear emphasis on quantitative (and easily measurable) network parameters like, e.g., packet loss rate, throughput, delay, jitter and/or bandwidth.

Of course, this reduction of research scope is strongly based on the implicit assumption that improving the quality of packet delivery with respect to one or more of these parameters will automatically lead to some sort of increase in user satisfaction. While this assumption in general is not unreasonable, we have to note on the other side that reality sometimes may be much more complex. Let us, for instance, consider the user-perceived quality of file download in a mobile broadband scenario, where downlink bandwidth can be safely assumed to play a key role as QoS parameter. Depending on the complexity of the task, it has been experimentally shown that users

A.M. Hadjiantonis and B. Stiller (Eds.): Telecommunication Economics, LNCS 7216, pp. 158–166, 2012.

demonstrate two different types of behaviour: as long as they are performing very simple download tasks (e.g. download of an mp3 or zip file), it seems that they are simply evaluating their waiting time until task completion (i.e. file download completed), leading to a logarithmic law as expected from what we know already from psychology [1].

However, if it comes to web browsing, this indirect proportional relationship between downlink bandwidth and user waiting time (i.e. doubling the bandwidth reduces waiting time by 50%) is no longer valid, basically for two reasons [1]: Firstly, due to complex interactions of the HTTP and TCP protocols with network performance, the network-level page load time for web pages does not directly depend on the available bandwidth. Moreover, rendering and displaying the web page on the local machine leads to additional non-linearities also for the resulting application-level page load time. Secondly, there is also a noticeable difference between perceived subjective page load time and the page load time on the application level, caused by the simple fact that while browsing through web pages, users regularly perceive the load process of a web page as already finished while in reality content is still being retrieved (e.g. because the browser window may be too small to display the web page in its entirety, or because the user, due to progressive rendering of the browser, does not anticipate that there might be additional content still under way). Indeed, experimental results indicate that the technical page load time typically differs from the perceived page load time by a surprisingly high factor between 1.5 and 3 [1].

Therefore, recently the strict user-centred emphasis of the original QoS concept has been reinforced, underlined by the new terminology of QoE whose fundamental definition again is due to ITU-T and reads as "*overall acceptability of a service or application, as perceived subjectively by the end-user*" [6]. Note, however, that the research community is still far from agreeing on this definition – among the various attempts to improve it, maybe the proposal elaborated in 2009 during a related Dagstuhl seminar is most noteworthy, describing QoE as "*degree of delight of the user of a service, influenced by content, network, device, application, user expectations and goals, and context of use*" [2].

2 Microeconomic Service Valuation

Indeed, we can assume an even broader interdisciplinary perspective while discussing appropriate concepts for determining the value of a communication service, leading us directly into microeconomic utility theory, which aims at describing the preferences of user i with the help of a so-called "utility function" $u_i(x)$. To this end, we formally define $u_i(x)$ to be a mapping of the consumption set X, i.e. the set of all resources user i could possibly consume, to real numbers such that $u_i(x) \leq u_i(y)$ implies that the user prefers y over x, with $x, y \in X$.

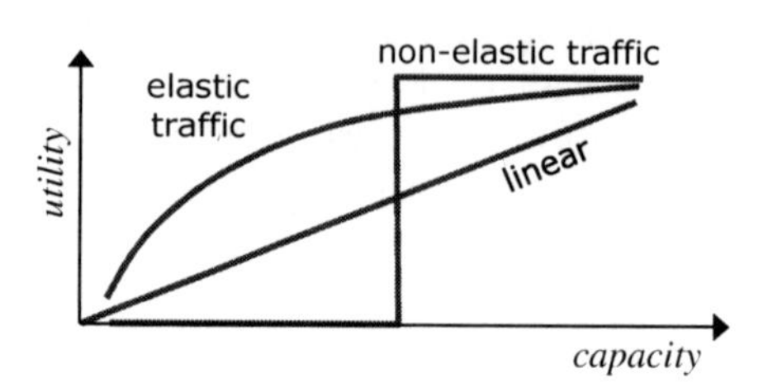

Fig. 1. Typical examples of utility functions

Figure 1 displays some typical shapes of utility functions for the case of network capacity: while linear utility assumes a direct proportionality between capacity and user satisfaction, the non-elastic case refers to applications (e.g. certain codecs) which require a certain minimum capacity but do not gain from additional capacity. Probably the most typical case is referred to as "elastic traffic", where initially some small capacity has quite a positive effect, whereas the impact of additional capacity is decreasing along with the level of already offered capacity (formally speaking, this mapping is assumed to be monotonically increasing, continuous - or at least right-continuous - and concave).

Amongst the many examples, weighted logarithmic utility functions of the form $u_i(x_i) = w_i \log x_i$ have been traditionally playing a key role in illustrating this concept. Note that in this case, if we assume a total number of users N, i.e., $i \in \{1,...,N\}$, the overall social welfare, defined as sum of user utilities, equals

$$U(x_1,...,x_N) = \sum_i w_i \log x_i \tag{5.1.1}$$

As has been shown by Kelly [5], the flow allocation x_i^* which maximizes (5.1.1) under capacity constraints of the form $\sum x_i \le C_j$, where the sum is taken over all flows i with routes $r(i)$ using link j with its associated capacity C_j, fulfils the criterion of *weighted proportional fairness*, i.e. there is no other allocation x with positive sum of proportional rate changes:

$$\sum_{i:j\in r(i)} w_i \frac{x_i - x_i^*}{x_i^*} = \sum_{i:j\in r(i)} w_i \frac{\Delta x_i^*}{x_i^*} \le 0 \tag{5.1.2}$$

Hence, as a primary conclusion we note that optimizing the overall social welfare under the assumption of logarithmic utilities leads to a proportionally fair allocation of the available network bandwidth.

3 WQL Hypothesis and Weber-Fechner Law

While so far, the assumption of a logarithmic utility function has been made mainly for reasons of mathematical tractability, we will now discuss whether and to which extent this can be linked to fundamental laws governing the perception of service quality in real life. Indeed, there is a significant number of examples in the related literature which report on such logarithmic relationships between a certain networking parameter (used as trigger/stimulus) and the resulting Quality of Experience, usually expressed in terms of Mean Opinion Score (MOS) values, including scenarios like Voice over IP quality depending on varying bitrates as evaluated by Rubino et al.'s Pseudo-Subjective Quality Assessment (PSQA) tool [14], web browsing under IP latency, using an experimental design where IP latency can be assumed to be equivalent to user waiting time [3], download of files (different sizes) or pictures in mobile broadband scenarios [1], and connection setup times for accessing a 3G mobile network [1].

It turns out that especially for simple tasks like "download this file", "connect to this mobile network", "start a Google search on the term 'xyz'", "go to the next picture in the gallery", etc., we observe a fundamental logarithmic law for QoE evaluation throughout. This has been recently termed "WQL hypothesis", stating that "*the relationship between **W**aiting time and its **Q**oE evaluation on a linear Absolute Category Rating (ACR) scale is **L**ogarithmic*". As argued in [1], there is sufficient empirical evidence suggesting that the WQL hypothesis cannot be rejected.

Should we be surprised by this result of logarithmic laws governing our perception of service quality at least to some extent? The answer to this question leads us even much deeper into the interdisciplinary arena, i.e. to the field of psychophysics as the science of quantifying the general behaviour of the human sensory system. The establishment of psychophysics dates back well into the middle of the 19th century when the German physiologists Ernst Heinrich Weber and Gustav Fechner first described what soon should become a very fundamental contribution to psychology of perception, i.e. the so-called Weber-Fechner Law. Their theory is based on the key concept of "just noticeable differences" which are assumed to be at the core of the human sensory system. According to this principle, sensory differences can be observed only if the corresponding trigger (physical stimulus) is changed by at least a certain proportion of its current value. Formally speaking, the differential perception dP is assumed to be directly proportional to the relative change dS/S of a physical stimulus of magnitude S. Assuming k as constant of proportion, straightforward integration then yields

$$dP = k \cdot \frac{dS}{S} \quad \Rightarrow \quad P = k \cdot \ln \frac{S}{S_0} \qquad (5.1.3)$$

where P describes the magnitude of perception as a function of the stimulus size S and a stimulus threshold (constant of integration) S_0.

The Weber-Fechner Law (WFL) can be applied to a surprisingly broad range of scenarios, ranging from vision (logarithmic stellar magnitudes), hearing (logarithmic dB scale), tasting, smelling, touching etc. However, its validity is not only restricted to actual human senses, but includes as well numerical cognition [8] and time perception [15].

4 The Fixed Point Problem of QoE Charging

Summarizing briefly what we have discussed so far, we have identified fundamental logarithmic dependencies between certain network parameters (QoS parameters) serving as stimulus and the user-centric perceptional evaluation of the resulting service quality for a broad variety of scenarios and evaluation methods (ranging from standards over learning tools to actual user trials). Moreover, microeconomic theory suggests that this kind of logarithmic law has further very interesting properties, most notably the fact that distributing network resources in a (proportionally) fair manner leads to the maximization of overall user benefit (social welfare). Bringing both lines of argumentation together, we have thus been able to show that indeed proportional fairness of resource allocation is equivalent to maximizing overall Quality of Experience – a relationship which to the best of our knowledge has not yet been formulated expressively so far.

While we started this section with a formal introduction of the utility function $u_i(x)$ as a user-specific preference function, there are different approaches if it comes to quantifying $u_i(x)$, e.g. by estimating the reselling value of a resource, or the willingness-to-pay of user i. Sticking to this latter approach, $u_i(x_i) = w_i \log x_i$ can hence be interpreted as the maximal charge/tariff user i would be willing to pay for the delivered QoS (e.g. bandwidth) x_i.

In this way, we have a basic mechanism to charge for QoS – determine the actually delivered QoS parameter, apply a suitable tariff function and calculate the resulting price. In this simple model, there is a primary (network level) feedback cycle as the charged price triggers the overall demand which itself is the key factor for the degree of congestion in the network and thus influences the provided Quality of Service (see Fig. 1 left).

This process is fundamentally different for the case of QoE charging (see Fig. 2 right): on the one hand, here we still pay for quality (now: Quality of Experience), i.e. we have to provide an estimation of the delivered QoE and calculate a charge from that. What makes life complicated, however, is the fact that the QoE estimation may heavily depend on the expected price itself (as well as on the QoS delivered by the network and also other factors). In this sense, the charge is serving both as input and as output of the QoE evaluation, thus constituting a secondary (user level) feedback cycle: high prices let the QoE expectations grow, hence the actual QoE evaluation will deliver relatively low results which by itself cannot justify the initial high prices. Similarly, low prices do not constitute major expectations, and as a consequence, the user is positively surprised by the experienced quality and would even be willing to pay more for it than what is actually charged.

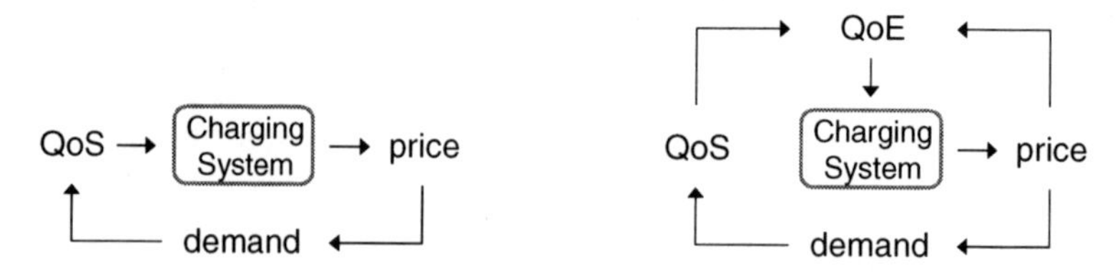

Fig. 2. Charging for QoS versus Charging for QoE

It is pretty obvious that this user-level feedback leads us to a very general fixed point problem – determining the precise charge which causes the right amount of expectations such that the subsequent QoE evaluation leads to a result for which the user is willing to pay exactly the amount she is asked for from the very beginning. While the determination of willingness to pay for certain perceived quality has experienced a comparably broad treatment in the related work, this is not the case for the other side of the medal, i.e. the influence of user predisposition caused by knowledge about the tariff structure onto her evaluation of perceived quality. To the best of our knowledge, the only experimental evidence so far is due to a series of user trials conducted in 2000/2001 as part of the European FP5 project M3I (Market-Managed Multiservice Internet).

Here, the idea was to offer users short video clips delivered with different frame rates (ranging from 1/sec to 25/sec) and ask them to indicate by a slider their willingness to pay. In addition, before starting the trial each user has been informed that she is member of one out of three categories (gold/silver/bronze) with VIP treatment at high charges for the gold users, preferential treatment at medium charges for the silver users and ordinary treatment at low charges for bronze users. Note that, actually, during the experiment no user differentiation whatsoever has occurred. For further details on the trial setup we refer to [3], which also provides the source for Figure 3.

From these results, we observe that – independently of the level of quality delivered – user expectations have indeed some influence on the quality evaluation. As far as acceptability is concerned (Fig. 3 top), gold users, who have been suggested to experience preferential ("VIP") treatment, show significantly higher expectations compared to silver/bronze users, while, at the same time, they are also willing to pay significantly more than their non-VIP colleagues (Fig. 3 bottom). Hence we may conclude that the user predisposition has indeed significant influence on her QoE evaluation, thus confirming the need for a detailed analysis of the mentioned secondary feedback cycle.

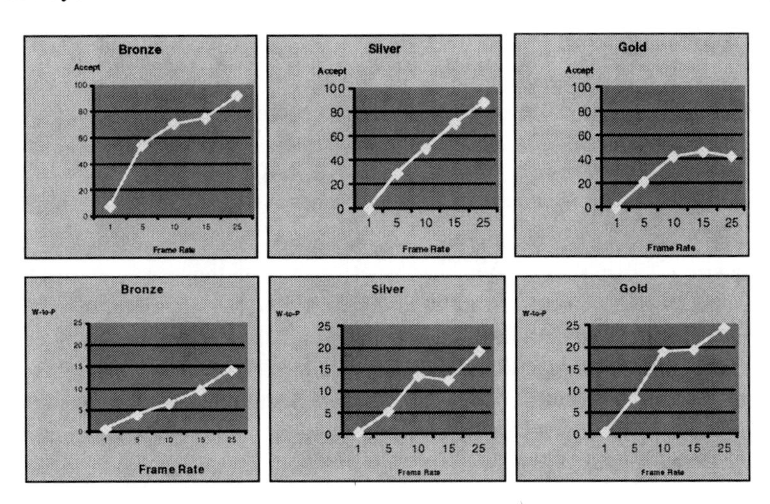

Fig. 3. Acceptability (in percent, first row) and Willingness-to-Pay (in pence per min, second row) for Different User Categories (source: [3])

5 Mechanisms for QoE-Based Charging

Depending on the location where the fixed point problem introduced in the previous section is supposed to be solved, we may distinguish two main approaches for QoE-based charging mechanisms: either, the transformation from QoS measurements to QoE estimations is performed by the system using an appropriate tool, while the result serves as input into a conventional Charging and Accounting System (CAS), or we leave this difficult task entirely to the end user who is put in charge to decide in real-time on his willingness-to-pay for the currently experienced service quality.

Whereas the first approach basically requires extending the CAS by some automatic QoE evaluation mechanism, e.g. based on learning algorithms (PSQA [14] and similar tools could provide invaluable help in this respect), we consider the second option to be by far more interesting. For instance in [10] we have proposed a feedback-based approach which is reducing the input required from the end user to a single bit of information indicating that both (a) the current QoE is unsatisfactory and (b) the end user is willing to pay for better quality. In terms of technical implementation, this solution only requires the installation of a "hot button" to be pressed by the end customer if both (a) and (b) are fulfilled, and as long as the additional quality (at the additional price) is in equilibrium with the user's expectations and needs. For further details about the prototypical realization of this concept in the framework of an IMS test-bed we refer to [10].

6 Toward the Future Telecommunications Ecosystem

This section has been devoted to exploring research questions at the precise intersection of microeconomics, psychology of perception and networking technology, demonstrating how such an interdisciplinary approach is able to make us aware of and to open us to a plethora of links and bridges between these disciplines which we initially had hardly hoped for. At the same time, we consider this as a very typical feature for the way telecommunications research will have to be conducted in the future in order to be able to contribute significantly to the sustainable success of this industry as a whole. From this point of view, it is no longer sufficient to restrict our research to more or less pure communications engineering, but on the contrary it is essential to continuously integrate the economic and user perspective as well.

This holistic approach requires at the same time a paradigm change which can be best described by the transition from communication systems to *communication ecosystems* (see [8] for a pivotal contribution to the establishment of this new overarching framework notion). Remember that the concept of an "ecosystem" as we know it from biology has turned out very useful in describing a community of organisms together with their environment, viewed as a system of interaction and interdependent relationships between the inhabitants as well as with the environment. It is most remarkable that the typical structure of a biological ecosystem, viewed as a pyramid composed of several layers (inorganic matter, basic source of food, primary consumers, secondary consumers and tertiary consumers = carnivores), which describe the most general interaction as of "eating or being eaten", strikingly resembles the classical layered structure of communication systems (physical layer, link layer, network and transport layer up to application layer) as we know it for instance from the ISO/OSI model. However, we now realize that the ISO/OSI approach only describes the environmental (= network technology) part of our communication ecosystem, whereas the holistic integration of its inhabitants, be it end costumers, business units or network, service, application, content providers, will require a significant extension way beyond layer 7. As a first consequence, this novel approach does no longer focus technology evolution as such, but rather on investigating how we can use it and what we can do with it.

Of course, putting such a holistic and interdisciplinary approach into practice is a different (and equally interesting) story. To this end, already almost a decade ago we have proposed to describe this interrelation between Network efficiency (economics), User acceptance (usability/user-perceived quality) and Technological feasibility (technology) as a kind of dynamic triangle of forces which we used to call "NUT Trilemma" [11]. However, it is essential to acknowledge that, while we believe that technology as such can no longer serve as the ultimate goal of our work, starting from solid technological grounds is still considered the indispensable first step which allows for the subsequent integration of economic and user aspects into a holistic framework.

Acknowledgments. Part of this work has been funded by the Austrian government and the City of Vienna within the COMET funding program. Additional support from COST Action IS0605, SISCOM/UEB (in the framework of the International SISCOM Research Chair 2010/11), INRIA Rennes and LAAS-CNRS Toulouse is gratefully acknowledged.

References

1. Egger, S., Reichl, P., Hoßfeld, T., Schatz, R.: Time is Bandwidth? Narrowing the Gap between Subjective Time Perception and Quality of Experience. In: Proc. IEEE International Conference on Communications (ICC 2012) – Communication QoS, Reliability and Modeling Symposium, Ottawa, Canada (June 2012)
2. Fiedler, M., Kilkki, K., Reichl, P.: From quality of service to quality of experience. Dagstuhl Seminar Proceedings 09192, `http://drops.dagstuhl.de/opus/volltexte/2009/2235/pdf/09192`
3. Hands, D., Gale, C., De Bruine, A.: User Experiments and Trials. FP5 project M3I, Deliverable 15.2 (November 2001)
4. Ibarrola, E., Liberal, F., Taboada, I., Ortega, R.: Web QoE evaluation in multi-agent networks: validation of ITU-T G.1030. In: Proc. ICAS 2009, Valencia, Spain (April 2009)
5. ITU-T Rec. E.800: Terms and definitions related to quality of service and network performance including dependability (1994)
6. ITU-T Rec. G.100/P.10: Vocabulary for performance and quality of service, amendment 2: new definitions for inclusion in Recommendation P.10/G.100 (2008)

7. Kelly, F., Maulloo, A., Tan, D.: Rate Control in Communication Networks: Shadow Prices, Proportional Fairness and Stability. Journal of the Operations Research Society 49, 237–252 (1998)
8. Kilkki, K.: Quality of Experience in Communications Ecosystems. Journal of Universal Computer Science 14(5), 615–624 (2008)
9. Longo, M.R., Lourenco, S.F.: Spatial attention and the mental number line: evidence for characteristic biases and compression. Neuropsychologia 45, 1400–1406 (2007)
10. Reichl, P., Fabini, J., Kurtansky, P., Stiller, B.: A Stimulus-Response Mechanism for Charging Enhanced Quality-of-User Experience in Next Generation All-IP Networks. In: Proc. XIII Conferencia Latino-Ibero-Americana de Investigación de Operaciones (CLAIO 2006), Montevideo, Uruguay (November 2006)
11. Reichl, P., Hausheer, D., Stiller, B.: The Cumulus Pricing Model as an Adaptive Framework for Feasible, Efficient and User-friendly Tariffing of Internet Services. J. Computer Networks (2003)
12. Reichl, P., Egger, S., Schatz, R., D'Alconzo, A.: The Logarithmic Nature of QoE and the Role of the Weber-Fechner Law in QoE Assessment. In: Proceedings of the 2010 IEEE International Conference on Communications, pp. 1–5 (May 2010)
13. Reichl, P., Tuffin, B., Schatz, R.: Logarithmic Laws in Service Quality Perception: Where Microeconomics Meets Psychophysics and Quality of Experience. Telecommunication Systems Journal 55(1), June 18 (2011), doi:10.1007/s11235-011-9503-7 (to appear, January 2014)
14. Rubino, G.: Quantifying the Quality of Audio and Video Transmissions over the Internet: the PSQA Approach. In: Barria, J. (ed.) Design and Operations of Communication Networks: A Review of Wired and Wireless Modelling and Management Challenges. Imperial College Press (2005)
15. Takahashi, T.: Time-estimation error following Weber-Fechner law explain subadditive time-discounting. Medical Hypotheses 67(6), 1372–1374 (2006)
16. Weber, E.H.: De Pulsu, Resorptione, Auditu Et Tactu. Annotationes Anatomicae Et Physiologicae. Koehler, Leipzig (1834)

Decision Support in Contract Formation for Commercial Electronic Services with International Connection

Martin Waldburger and Burkhard Stiller

University of Zürich, Communication Systems Group (CSG), Switzerland
{waldburger,stiller}@ifi.uzh.ch

Abstract. This section addresses the automated and legally compliant determination of key contract parameters for commercially and internationally provided electronic services in the Internet. It designs and implements a decision support system that provides recommendations on suited contract parameters at the time of contract formation. Recommendations base on automated reasoning taking into consideration the relevant set of service- and contract-specific facts. The range of facts to be considered is imposed by the laws of these states that have sufficient connection with the international service contract to be concluded. Accordingly, this section develops and documents the solid basis for the decision support system, consisting of an information model that integrates perspectives of service and contract management as well as in a detailed method to formally model the legal basis for machine execution.

Keywords: Decision support, electronic service, Private International Law (PIL), contract, dispute resolution.

1 Introduction and Motivation

Electronic business in the Internet has become an important driver for economic growth. The provisioning of commercially offered electronic services in the Internet — *e.g.*, content, news, or social networking services — requires the conclusion of an international contract in case the respective service is provided across borders. In international contracts, two contractual parameters are of key importance: jurisdiction and applicable law. Jurisdiction indicates which state's courts are authorized to hear and decide on a potential contract conflict [1], while applicable law indicates under which state's law a court decision shall be found [2].

The way jurisdiction and applicable law choices are made in international service contracts today is often not compliant with the relevant provisions of Private International Law (PIL). Jurisdiction and applicable law provisions are usually present in terms of choices made, but these choices may be illegitimate. Illegitimate choices are voided (and replaced by PIL-compliant terms) should a dispute arise and a contract claim be deposited in a court. Jurisdiction and applicable law provisions, thus, might have a considerable impact on the risk assessment of an international service contract to be concluded [3–5]. Given this risk and the imminent uncertainty outlined, service providers

A.M. Hadjiantonis and B. Stiller (Eds.): Telecommunication Economics, LNCS 7216, pp. 167–178, 2012.

and customers alike need support in forming international service contracts [5, 4, 6, 7]. In particular, they need to know about jurisdiction and applicable law choices they can rely on so that these parameters may become part of an SLA (Service Level Agreement). To date, however, there is no alternative available to the static, PIL-ignorant way adopted currently. This lack is perceived as a major hurdle to foster adoption of (international) electronic business.

Hence, this section[1] is a pioneering effort to support service providers and service customers in international service contracting by means of a decision support system developed [8, 9]. This system produces a list of recommended jurisdictions and/or applicable laws during contract formation phase. Recommendations are determined in an automated and compliant manner according to the PIL-driven contract- and service-specific set of connecting factors. Hence, it forms a dedicated task of Service Level Management (SLM).

This implies a number of challenges to be addressed, as there is considerable complexity in selecting the right PIL(s), modeling the accordingly relevant provisions, and implementing modeled laws in terms of a decision support system to produce jurisdiction and applicable law recommendations. In order to reflect and integrate different notions originating from different jurisdictions and their laws, a common information model basis is built. In the light of a method lacking to identify, select, and formally model the relevant legal basis, such a method is developed. In consideration of both, modeling method and information model, an implementation method is determined. Finally, an automated determination of jurisdiction recommendations is shown feasible and fully operational for the example of the main European PIL regulation modeled and implemented.

2 Background Information and Related Work

Given the inherent complexity of procedures in PIL the following gaps in concepts, models, and implementation have to be met:

- *Information model, modeling method, and implementation*: The analysis of state-of-the-art in those areas of scope considered has revealed that there are substantial gaps (a) in a common information model considering dimensions of service and contract management, (b) in a lacking method to identify, analyze, and formally model a PIL of interest, and (c) in a decision support system implementation helping contract parties by means of reliable recommendations on jurisdiction and applicable law. These three components (a) to (c), hence, constitute the key set of major contributions focused in this section.
- *Applicable notion of service and contract*: The modeling method, the implementation, and especially the information model shall consider and reflect concepts as well as concrete information artifacts in relation to the type of service (purely electronic services for monetary compensation) and contract (bilateral international service contract under PIL) assumed here. Moreover, the contract type of a service

[1] Thesis available at `http://www.csg.uzh.ch/staff/waldburger/extern/publications/Dissertation-Martin-Waldburger.pdf`

contract shall be investigated and characterized in detail. The comprehensive analysis and emulated contract qualification of a service contract as performed in a separate case study shall provide for a fourth major contribution of this section in addition to the information model, modeling method, and implementation.

- *Consistency and compliance*: In particular the modeling method, but also the information model and the implementation, shall consider and reflect characteristics of PIL procedures. This implies, for example, a time-wise back porting of dispute-dependent PIL provisions to the time of contract conclusion. Overall, legal compliance and content-wise consistency with a law to be modeled are key issues while interpretation, albeit not completely avoidable, shall be kept to a minimum.
- *Extension of an existing information model*: The information model provides the common set of information concepts and information artifacts for both, modeling method and implementation. The SLM information model as developed and documented in [10] qualifies as a well-suited, established basis for model extensions. Model adaptations are mainly needed to reflect both, service and contract management dimensions.
- *Rule-based system*: The implementation shall adopt a rule-based system-driven approach, as rule-based systems show advantages in building expert systems addressing decision-based procedures that are characterized by high complexity. Hence, with the help of a knowledge base, a modeling result consisting mainly of conditions and actions, and the inference engine, a decision support system as introduced and motivated shall be implemented in logic programming.

3 Research Methodology

Figure 1 outlines the overall approach developed within this section. The first step consists in identifying potentially affected jurisdictions by an international service contract to be concluded. This should happen in the same way a court dealing with a PIL-oriented claim would proceed. A court would collect basic connecting factors and determine on this basis jurisdictions with potential connection. In the contract conclusion case, such procedure is to be reflected by the contract parties to submit the respective set of contract party- or service-specific connecting factors of interest. A complete implementation takes these factors in consideration and produces a list of supposedly connected jurisdictions. For each jurisdiction identified, the set of relevant PIL sources is determined. Criteria for the PIL selection are related to application of a PIL in question, namely whether a law applies to a case in question (material application), whether it is in force for the time frame in question (temporal application), whether it applies in the location or locations touched (geographical application), and whether it supersedes other PILs or is subsidiary to another PIL (hierarchical application).

In step 2, each PIL identified needs to be reflected by a formal model (*e.g.*, in terms of an activity diagram; one diagram per PIL), which is implemented in order to produce jurisdiction/applicable law-related output. Modeling, implementation, and output generation base on a common information model as well as a common modeling and implementation method. By every new PIL modeled, common parts might need to be altered in order to reflect so far non-covered aspects. Updates in the information model

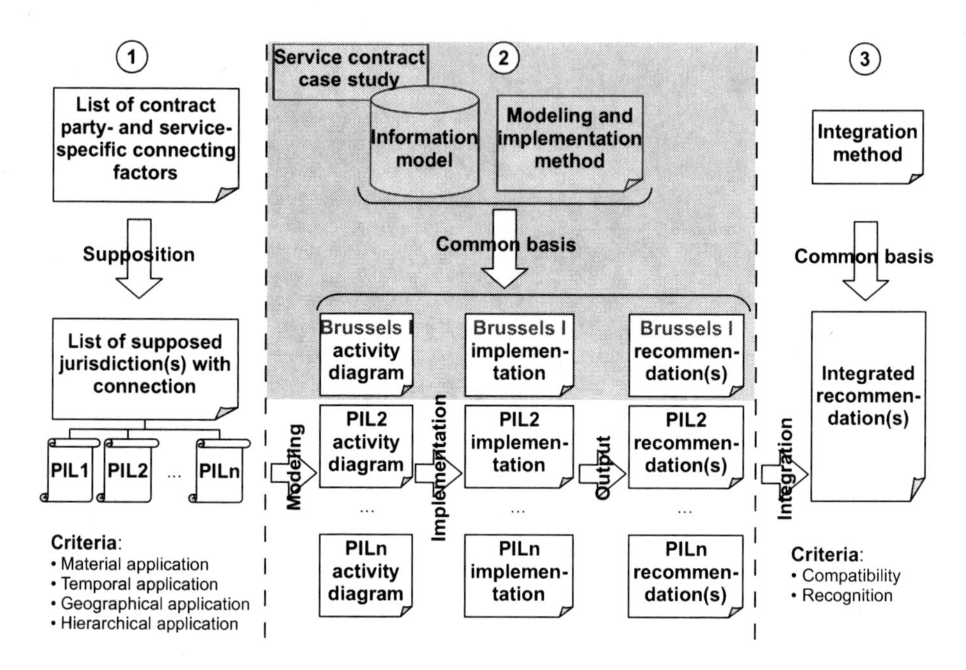

Fig. 1. Overall Three-step Approach and Focus Area (Gray Background)

and the method are expected to tend less frequent with the number of already modeled PILs. Moreover, changes in the underlying information model might provoke an update in the set of basic connecting factors to be collected during step 1. Connecting factors are of utter importance since if an important connecting factor is missing in the first place, a relevant jurisdiction (and related PILs) might not be investigated at all.

A list — one list per PIL considered — of recommended jurisdiction(s) and/or applicable law(s) is produced in step 2. These lists constitute the main outcome of this step. The procedure, however, cannot end here. Contract parties should obtain additional information with respect to compatibility of the different recommendations. Due to a territorial principle in law (state sovereignty) there is no mechanism in place, *per se*, that guarantees consistency (in terms of compatibility) within jurisdiction/applicable law provisions originating from different PILs. Only in cases where a single list of recommendations is produced — a case, however, which is rather unlikely as it is a very special case — or in case multiple lists are fully compatible (congruent lists) the mitigation strategy to adopt is clear.

The focus of this section is exclusively on the core of step 2, including several substeps (*cf.* area with gray background in Figure 1). In particular, a thorough investigation of how to formally reflect a PIL leads to substantial methodological contributions. Methodology application is assessed for an example PIL modeled, the Brussels I [11] regulation. The detailed modeling method with respect to PIL identification, PIL selection, and law analysis is developed accordingly. Focus in this context is put on major cases meaning that the law analysis method includes aspects assumed to reflect typical circumstances while aspects such as reservations are not considered. The functional modeling method is completed by means of guidelines and criteria outlined for a successful PIL modeling.

In order to develop a comprehensive understanding of international service contracts, this section's fourth major contribution beyond information model, modeling method, and implementation, is found in an extensive case study to conduct a simulated contract characterization. The accordingly developed notion of an international service contract determines a key driver to shape a suited information model. The type of contract investigated is in relation to Bandwidth-on-Demand (BoD), while the contract qualification itself is simulated by means of nominate contract types of the Swiss code of obligations [12].

The investigation in a BoD product case is conducted by means of a real-world BoD product offering. The BoD product family looked at in detail is termed Ethernet Virtual Private Line (EVPL) [13] as offered by Verizon Business. This BoD product case embraces the Verizon EVPL-Metro and -National BoD offerings. More specifically, it includes the embracing study of the complete set of customer agreement components, covering service guides, terms and conditions, charge definitions, terminology definitions, and SLAs.

3.1 Design Science Approach

Figure 2 visualizes the research methodology in relation to modeling and implementation phases. The methodology follows a design science approach (as opposed to a behaviorist approach) [14] and is represented based on the method introduced by [15]. Accordingly, this section is mainly concerned with the design of artifacts and the contribution of scientific knowledge. In terms of the research object envisaged, a socio-technical focus on international service contract-related, electronic business-related transactions is taken. In this context, action system and information system artifact development is envisioned. The first reflects the work flows interpreted from PIL procedures to determine jurisdiction and applicable law. The second reflects the actual implementation in terms of a system to enable an automated jurisdiction and applicable law recommendation determination for a service provider and a service customer at the time of contract formation.

In addition to action and information system artifacts, the key contribution in terms of knowledge contribution consists, on the one hand, in the underlying design artifact developments of the respective information and work flow models, on the other hand, in a conceptual framework how to identify, analyze, and model multiple PILs as well as related service types in a technically and legally correct, economically efficient, and scalable manner. The information model and work flow model find expression in a formal language, namely in UML2 Activity Diagrams and UML2 Class Diagrams (information model) as well as formally expressed rules. The modeling and implementation method, *i.e.*, the applicable conceptual framework, finds expression in natural language.

The overall methodology bases on a central hypothesis and it follows a common purpose. The hypothesis adopted is driven by a service provider's and service customer's identified need for improved risk assessment along the complete contract and service life cycle with respect to international service contracting. The automated and legally compliant determination of jurisdiction and applicable law recommendations for

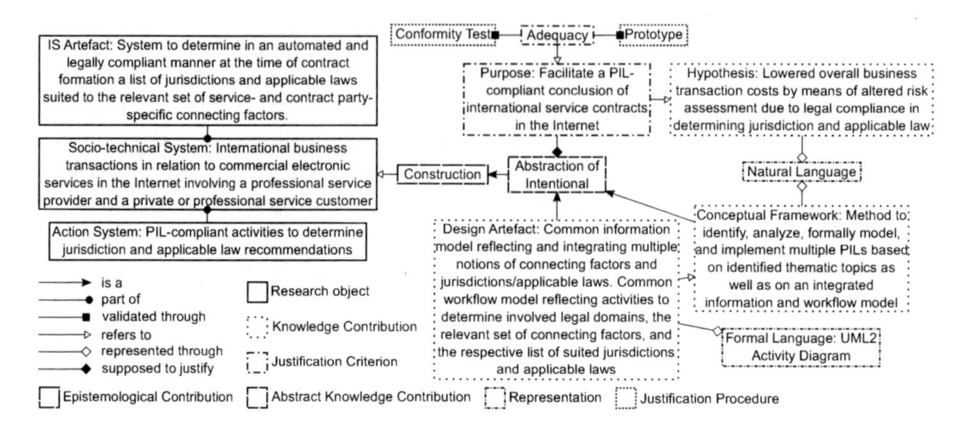

Fig. 2. Research Methodology

international service contracts shall reveal increased predictability and legal certainty for both contract parties.

4 Achievements

With the relevant purpose, hypothesis, and problem formulated, the set of claims have been translated into the key set of five specific objectives with respect to the design, implementation, and testing of a decision support system to produce jurisdiction and/or applicable law recommendations. Claims, objectives, and primarily the identified gaps have shaped the range of four major contributions as follows:

- *International service contract case study*: Based on a real-world, complex service contract construct assessed, a profound service and international service contract understanding has been obtained.
- *Information model covering service and contract management*: The understanding gained in the case study conducted has provided the solid basis for extending an established information model in SLM with the relevant perspective on contract management. In particular, information concepts and information artifacts have been determined to reflect characteristics of international service contracts.
- *PIL modeling method*: In order to ensure legal compliance and consistency with the relevant PIL basis an extensive method to identify, analyze, and formally model a PIL of interest has been developed. This method considers information concepts and artifacts of the information model determined. It results in a UML2 activity diagram representing a given PIL source.
- *Decision support system producing recommendations*: Driven by the outlined problem to be solved in this section, a service provider and a service customer need decision support when forming a service contract with international connection. To this aim, a previously modeled activity diagram is implemented as a rule-based system establishing a decision support system which produces recommendations on jurisdiction(s) and/or applicable law(s).

In terms of gaps addressed, a detailed results valuation discussion has revealed that all four contributions have been fully and successfully achieved. Equally, all five gaps identified were found completely addressed.

4.1 Claims and Objectives Addressed

In consideration of those four major contributions achieved and the set of five gaps successfully addressed, four claims are assessed qualitatively as follows:

- *Trust-building in international electronic business*: Lack of trust in international electronic business is perceived as a major hurdle to wider adoption of commercial electronic service provisioning and to implement harmonized markets. Accordingly, this section aims to foster trust of service customers and service providers in international electronic business. The existence of additional information at the time of contract formation made available by the decision support system implemented is clearly rated as a trust-building measure. Additional information shows characteristics of recommendations. Without the decision support system, contract parties do not dispose of any information at all. In particular, they lack a notion of suited, *i.e.*, recommendable jurisdiction(s) and applicable law(s) — recommendations that were determined according to the relevant legal basis, following a fully documented law modeling and implementation method.
- *Increased legal certainty in international service contracting*: The observed status quo in international service contracting bears considerable risks for both contract parties. With respect to dispute resolution means, this is mainly due to legal uncertainty in valuing risks of foreign jurisdiction and foreign law applied. None of the four major contributions achieved in this section is able to produce guaranteed legal certainty and to overcome the inherent design gap between the borderless infrastructure of the Internet and the territorial perspective of PIL. Nevertheless, this section reveals a path toward increased legal certainty, especially by the design and implementation of a decision support system which produces recommendations that both contract parties can rely on as they have been determined in accordance with PIL and based on a transparent method.
- *Legal compliance in determining jurisdiction and applicable law*: The production of a list of recommended jurisdiction(s) and applicable law(s) follows the procedures imposed by the respective PIL modeled and implemented directly. As long as the relevant set of PILs is modeled and implemented, the relevant set of recommendations may be obtained. Thus, legal compliance is provided, in principle. Alone, erroneously modeled provisions cannot be excluded completely. This is perceived less an issue of the PIL modeling method developed than an issue of how to correctly interpret running text in a law — laws do not constitute formal specifications. Hence, a certain degree of interpretation remains unavoidable.
- *Risk assessment due to informed choices and PIL awareness*: For producing a list of recommended jurisdiction(s) and applicable law(s), the respectively relevant connecting factors must be known. In case of Brussels I, the set of 12 connecting factors was required. In a productive decision support system, both contract parties would be required to submit a set of connecting factors initially. The fact alone that both

contract parties would have to answer imperatively about a number of these factors implies a minimum degree of awareness that the contract to be concluded would have international connection (*e.g.*, customer from a foreign state, provider from a foreign state). Given such minimum degree of awareness assumed, both contract parties are likely to realize that a potential dispute might bear certain risks — even if the actual detailed legal consequences may not be appreciable.

Such an approach fosters trust in international electronic business, increases legal certainty, determines recommendations in compliance with PIL, and allows for risk assessment due to informed choices and PIL awareness. Subsequently, the qualitative assessment of five objectives reads as follows:

- *Automation support*: The decision support system implemented allows for automated reasoning. It implements an activity diagram modeled for a given PIL as a rule-based system. Both, modeling method and decision support system implementation, envision an automated determination of jurisdiction and/or applicable law recommendations. Hence, automation support is provided, in principle. The way the decision support system prototype for Brussels I is implemented, however, relies on pre-determined input in terms of facts. Facts reflect connecting factors. In a productive decision support system, not only the reasoning, but also the acquisition of connecting factors from both contract parties would have to be implemented dynamically.
  The decision support system implemented is found to produce results in terms of recommendations. It produces recommendations in accordance with the underlying modeled and implemented PIL procedures considered, and it implements activity diagrams which result from a modeling method that translates any *ex post* provision (focus on contract enforcement phase) to the respective *ex ante* provision (focus on contract formation phase).
- *Integration of different national and supra-national PILs*: All contributions made throughout this section are designed in a way to allow for flexibility in terms of general application and extension. This means that, even though an example service contract and an example PIL was used to concretize contributions at all levels, the service and contract notion obtained in the case study conducted is generic enough that the developed information model features information concepts of wide applicability. With respect to information artifacts, the information model currently foresees specific artifacts for Brussels I and Rome I [16] regulations. Should further artifacts be needed when considering other PIL sources, additional artifacts may be added without the need to change existing artifacts, the existing Brussels I activity diagram, or the existing decision support system for Brussels I. Extension, thus, at all levels is possible and even foreseen. Furthermore, the section proves for the case of the Swiss national PIL that the PIL modeling method may be successfully applied to other PIL sources. The same holds true for an implementation of the respective decision support system as the implementation method is not specific to a single PIL, but it considers generally applicable guidelines to implement an activity diagram as a rule-based system.
- *B2C and B2B support*: The case study conducted considers a B2B contractual relation. All further contributions of this section consider B2C and B2B relations.

For instance, the information model includes a Brussels-I-specific information artifact holding information about whether a service customer is a consumer (B2C). Similarly, the activity diagram modeled for Brussels I and the according decision support system implemented differentiate B2C and B2B relations. In other PILs, this differentiation might not be relevant. Nonetheless, the information model, the PIL modeling method, and the decision support system implementation provide full support for both.

- *Scope of international service contracts*: The information model, the PIL modeling method, and the implementation method consider provisions of PIL relevant to international, commercial, and electronic provisioning of services exclusively. A wider focus including so far not envisaged service contracts of similar types may be accommodated at a later stage. The information model, the modeling method, and the implementation method support extensions beyond the scope adopted here.
- *Scalability by means of complexity reduction*: All contributions have revealed a considerable amount of complexity faced. In particular, the PIL modeling method is affected by complexity and scalability issues. Scalability with respect to a wider geographic, electronic business scheme, and service contract type coverage has been achieved mainly by means of complex reduction through abstraction. Consequently, the PIL modeling method focuses on standard cases. It emphasizes legal compliance and, thus, reliable recommendations produced for frequent and likely contractual relations. On the other hand, it abstracts exceptions and reservations. This decision reflects a trade-off between scalability and specifics. As the detailed inclusion and exclusion criteria for modeling a PIL source, however, are fully documented, any limitation in specifics is made transparent and comprehensible.

In summary, the qualitative analysis of those objectives raised reveals that all objectives have been met successfully. This means that this section provides sufficient support for automation and that it is extensible with respect to a wider geographic, business scheme, and service contract type scope. Extensibility and complexity reduction through abstraction contribute equally to achieve scalability in future work.

4.2 Future Work and Complexity Mitigation Strategies

The overall three-step approach as shown in Figure 1 has been developed according to procedures applicable to international contract claims. In this context, the contributions of this section determine major achievements in terms of a common basis for a PIL-specific, successful modeling, implementation, and output generation. On the other hand, wider extension in PIL coverage, supposition of potentially connected jurisdictions, and integration of different jurisdiction/applicable law recommendations have been identified as fields of interest for future work.

While the first may be achieved by making use of the modeling and implementation method as well as by the information model developed, supposition and integration are considerably more challenging, since both bear a high level of complexity and uncertainty. When aiming at a comprehensive, potentially even productive solution, however, the complete set of supposition (step 1 in Figure 1), PIL modeling, implementation, and output production (step 2), as well as integration of recommendations (step 3) has to be endorsed.

In general, the way complexity was addressed in developing the modeling method, the information model, and in the implementation method reflects a "divide-and-conquer" approach. In modeling, the overall procedure was split into distinct steps for which each either guidelines or criteria could be determined. For instance, law selection was split into a pre-selection and detailed in-/exclusion sub-procedure so that complexity was better handled.

In the information model, the starting point was found in an existing SLM model which was extended step-by-step. Again, this happened in order to accomplish a complex procedure. In the implementation, finally, a multi-step procedure focusing on different in- and output variables and predicates implementing partial paths (instead of complete paths) was adopted. The latter facilitates a partial debugging — reducing complexity considerably. These complexity issues lead to the question of how to mitigate such challenges. Three basic directions have been identified as follows:

- *Status quo*: One way to handle complexity is to not do anything *fundamental* about it, but to cope with it as good as it gets. Coping means to, *e.g.*, extend both, modeling method and PIL coverage range, possibly with the help of jurists in order to lower potential for misinterpretation. In fact, this strategy is perceived as a pragmatic albeit inefficient and probably only symptomatic approach. Given the fact that territoriality and the respective political implications that come with it are not expected to diminish any time soon, a status quo approach constitutes a realistic option.
- *Arbitration*: There are alternatives to judicial arbitration. Arbitration is a possible approach to so-called alternative dispute resolution. While alternative dispute resolution shows typically advantages in terms of flexibility and choice over judicial jurisdiction, it may as well turn out as complex as PIL, and enforcement might be a real problem in some cases. In conclusion, arbitration (and other alternative means) is difficult to assess with respect to trade-offs.
- *PIL for international service business*: The Internet is one of the few truly global infrastructures. Electronic business in the Internet is happening now, and so are international service contracts concluded every day. The existence of a harmonized, widely accepted PIL specific to international service contracts in the Internet would mark a corner stone towards making contracts in the Internet less of a "second-class" type of contract. Harmonization in PIL is perceived as the only way to address issues with the current approach at the root and in a sustainable manner. Therefore, it must be envisaged as a long-term objective.

5 Conclusions

This section has addressed an important problem in contract formation support, especially the automated determination of recommendable jurisdictions and/or applicable laws for an international service contract. Accordingly, the decision support system developed fosters transparency and trust in international electronic business. Due to the determination of jurisdiction and applicable law, the minimization of risks between service provider and service customer has been achieved. Thus, this section's major contribution to SLM and contract management tasks determines an easy-to-use and applicable functionality, which can be embedded into SLA management systems due to the availability of contract parameters and their values. Therefore, the practical considerations of such an approach for service providers and service customers are of high importance: (a) they are able to reduce their risks in contract formation and (b) new legislations and laws can be modeled according to the methodology developed and integrated into the existing implementation prototype.

Where minds of service providers, service customers, and policy makers might meet in the long run, is in an internationally harmonized PIL for commercially provided electronic services in the Internet. Such legislation, if drafted in consideration of the respective technical and legal requirements, may overcome those challenges raised by today's territorial approach to Internet jurisdiction. But it is safe to assume that any move in this direction will take time. Nonetheless, it is perceived the only way to essentially and fundamentally foster legal certainty in electronic business in the Internet. This conclusion is irrespective of the decision support system designed and implemented within the scope of this section. It would be of help in modeling and implementing decision support for both contract parties also if there was only a single PIL of relevance. The United Nations Convention on Contracts for the International Sale of Goods (CISG) [17] proves[2] that such an undertaking resulting in internationally harmonized law is feasible, in principle.

Acknowledgment. The authors would like to thank Stefan Bechtold, Thomas Schaaf, and Marinos Charalambides for their valuable input, comments, and feedback.

References

1. van Lith, H.: International Jurisdiction and Commercial Litigation. Uniform Rules for Contract Disputes. Asser Press (July 2009)
2. Kroes, Q.R.: E-business Law of the European Union. Kluwer Law International (2003)
3. Patrikios, A.: Resolution of Cross-border E-business Disputes by Arbitration Tribunals on the Basis of Transnational Substantive Rules of Law and E-business Usages: The Emergence of the Lex Informatica. University of Toledo Law Review 38(1), 271–306 (2006)
4. Geist, M.A.: Is There a There There? Towards Greater Certainty for Internet Jurisdiction. Berkeley Technology Law Journal, 1–63 (Fall 2001)
5. Wang, F.F.: Internet Jurisdiction and Choice of Law: Legal Practices in the EU, US and China. Cambridge University Press (2010)
6. Svantesson, D.J.B.: Private International Law and the Internet. Kluwer Law International (February 7, 2007)
7. Waldburger M., Macri A., Stiller B.: Service Provider Market Activities Constituting Jurisdiction for International Service Contracts—A Structuring Approach and Techno-legal Implications. In: 21st European Regional ITS Conference (ITS 2010), Copenhagen, Denmark, pp. 1–22 (September 2010)

[2] For international sales of material goods, not for electronic services.

8. Waldburger, M., Charalambides, M., Schaaf, T., Stiller, B.: Automated Determination of Jurisdiction and Applicable Law for International Service Contracts: Modeling Method, Information Model, and Implementation. In: 18th Biennial and Silver Anniversary International Telecommunications Society Conference (ITS 2010), Tokyo, Japan, pp. 1–31 (June 2010)
9. Waldburger, M., Stiller, B.: Automated Contract Formation for Electronic Value-added Services in the Internet—The Case of Bandwidth-on-Demand Contracts in Europe. In: 37th Research Conference on Communication, Information and Internet Policy (TPRC 2009), Arlington (VA), USA, pp. 1–30 (September 2009)
10. Schaaf, T.: IT-gestütztes Service-Level-Management: Anforderungen und Spezifikation einer Managementarchitektur. PhD thesis, LMU München, Fakultät für Mathematik, Informatik und Statistik, Munich, Germany (November 2008)
11. Council of the European Union: Council Regulation (EC) No 44/2001 of 22 December 2000 on Jurisdiction and the Recognition and Enforcement of Judgments in Civil and Commercial Matters. Official Journal of the European Communities L12, 1–23 (January 2001)
12. Bundesversammlung der Schweizerischen Eidgenossenschaft: Bundesgesetz vom 30. März 1911 betreffend die Ergänzung des Schweizerischen Zivilgesetzbuches (Fünfter Teil: Obligationenrecht), SR 220 (March 1911) (in German)
13. Verizon Business: Ethernet Virtual Private Line (EVPL) (May 2010)
14. Hevner, A.R., March, S.T., Park, J., Ram, S.: Design Science in Information Systems Research. MIS Quarterly 28(1), 75–105 (2004)
15. Frank U.: Towards a Pluralistic Conception of Research Methods in Information Systems Research. ICB-Research Report No. 7, Universität Duisburg-Essen, Essen, Germany (December 2006)
16. European Parliament and the Council of the European Union: Regulation (EC) No 593/2008 of the European Parliament and of the Council of 17 June 2008 on the Law Applicable to Contractual Obligations (Rome I). Official Journal of the European Union, L177, 6–16 (July 2008)
17. United Nations Commission on International Trade Law (UNCITRAL): United Nations Convention on Contracts for the International Sale of Goods (CISG). Convention (April 1980)

Competition among Telecommunication Providers

Patrick Maillé[1], Peter Reichl[2], and Bruno Tuffin[3]

[1] Télécom Bretagne, Cesson-Sévigné Cedex, France
[2] FTW Telecommunications Research Center Vienna, Austria
[3] INRIA Rennes Bretagne-Atlantique, Campus de Beaulieu, France
patrick.maille@telecom-bretagne.eu, reichl@ftw.at,
bruno.tuffin@inria.fr

Abstract. We analyze in this section game-theoretic models of competition between telecommunication networks providers in various contexts. This analysis helps to define and understand the operators' pricing and technology investments as well as the most efficient market rules by a regulator.

Keywords: Competition, game theory, pricing.

1 Context

The Internet has moved from an academic network to a commercial and highly competitive one where providers compete for customers by playing with the price to access service and the Quality of Service (QoS) they offer. There are now different ways to access the network through different technologies, using ADSL, telephone network, optic fiber, CDMA wireless networks, WiFi or WiMAX among others, with different QoS capabilities. Moreover, the convergence of networks, where Internet access, wired and wireless telephony, television are regrouped into a single network poses additional economic challenges. In this situation, each provider has to adapt his pricing scheme in order to attract customers, to maximize his revenue and/or to allow fairness in the way resources are shared. Pricing has therefore been a hot topic in telecommunication networks during the last decade (due to congestion) and many schemes have been proposed in the literature [1,2,15].

On the other hand competition among providers has received very little attention up to now. Though, telecommunication networks have become highly competitive and it seems primordial to us to deal with that competition in pricing models when defining the optimal prices, since competition may highly affect the results of price determination (while pricing in a monopolistic context generally means a single level of game between users, competition actually introduces an additional level of game, between providers). Some typical illustrations of competition are described below.

For wired access, DSL users can choose among several competing providers to connect to the Internet.

1. The case of wireless access is more flexible. For example a user wishing to connect to a WiFi hotspot may be located in a zone covered by several wireless access providers, and can choose which provider to use for the time of his connection.

A.M. Hadjiantonis and B. Stiller (Eds.): Telecommunication Economics, LNCS 7216, pp. 179–187, 2012.

The same user can/will even be able to choose between different and competitive transmission platforms: WiFi, WiMAX, 3G, wired operators, FTTx (Fiber To The X), with a possible combination of all those ones (the so-called multi-homing). Depending on the context, providers are or can be forced to share a part of the available resource.

Our goal in this section is to deal with that issue. Indeed, we need to study the distribution of customers among providers as a first level of game, and then to focus on a second higher level, the price and QoS war among providers. We aim at summarizing some of the main findings of the French ANR project CAPTURES (see http://captures.inria.fr/), run in conjunction with Econ@Tel, whose main goals are:

- To provide models and analysis of direct competition among providers operating on the same or different technologies (being WiFi, WiMAX, 3G, ADSL...) and the dynamics of those models.
- To discuss capacity expansion for providers from an economic point of view.
- To analyze retention policies that prevent users from churning, and how this can be controlled.
- In general, to discuss regulation rules that could help to drive the system to an optimal situation.

The key tools are telecommunication networks modeling, non-cooperative game theory and optimization.

Figure 1 illustrates the type of situation to be analyzed with different providers (DSL, wireless, etc.) not necessarily covering all users. Those providers have to define their price (p_i for provider i) at the largest time scale as a game (they can additionally play with capacity, QoS and other parameters). At the smallest time sale, depending on the offered QoS and the price at each provider, demand is split among providers. Those two games are strongly dependent: the way demand is distributed depends on prices set by all providers; similarly the price strategy of a provider will depend on demand distribution. What complicates more, even at the first step, demand depends on QoS, itself depending on the number of attracted customers. But we assume that providers play smartly by anticipating the reaction of users to the price profile (the so-called backward induction).

This type of two levels game is called a Stackelberg game [14], where providers are leaders playing first and anticipating the reactions of users.

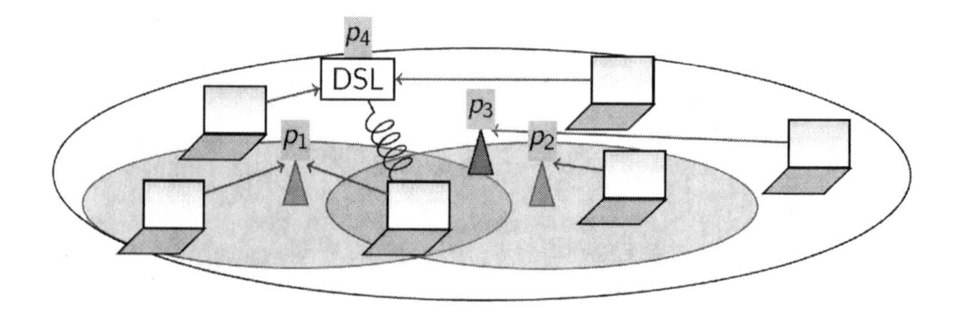

Fig. 1. Typical two-levels model

2 Demand Distribution among Providers and Influence on Price War

To model the competition among providers, we use the above two-levels Stackelberg game. The questions are: How to characterize the equilibrium in the distribution of customers among providers? Does it exist and is it unique? Is there a Nash equilibrium in the price war between providers? Recall that a Nash equilibrium is a vector of providers' strategies such that no provider can improve his revenue by unilaterally changing his own strategy. If it exists, is this equilibrium unique and what can drive the system to a "good" one?

There are several ways to represent the demand of end users, such as attraction models, coming from marketing theory (a well-known class being the generalized MultiNomial Logit (MNL) [3]. But for illustration purposes here, we consider a demand repartition coming from transport theory. In that case, each packet is like a car on a road, and its influence on the resulting traffic can be seen as negligible. Then demand will be split in such a way that for each route (at each provider), the perceived cost, made of the charge imposed on customers and the level of QoS provided, will be equal for each provider receiving some traffic. Indeed, otherwise part of the traffic at an "expensive" provider could be re-routed to a "cheaper" provider. The resulting equilibrium is called a Wardrop equilibrium [16].

We start with the model and results described in [10], where N providers compete on the exact same domain, as depicted in Figure 2. Time is slotted and Provider i is able to serve C_i packets (or units) per slot. If his demand is d_i, packets in excess (chosen uniformly) are lost and have to be resent. Figure 3 describes this loss model, the (uniform) probability of transmission success and the expected number trials before success (from a geometric distribution). Following [13], we assume that the price p_i at provider i is per sent packet and not received ones, a congestion pricing to incentivize users to efficiently use the capacity. The perceived price per received packet is then $P_i = p_i \max(1, d_i/C_i)$.

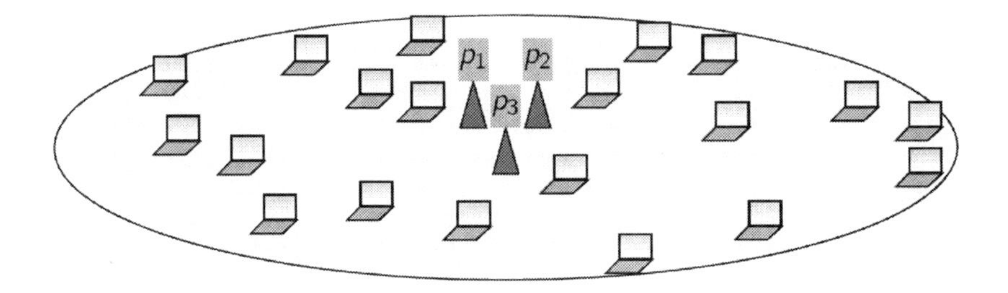

Fig. 2. Example with 3 providers covering the same domain

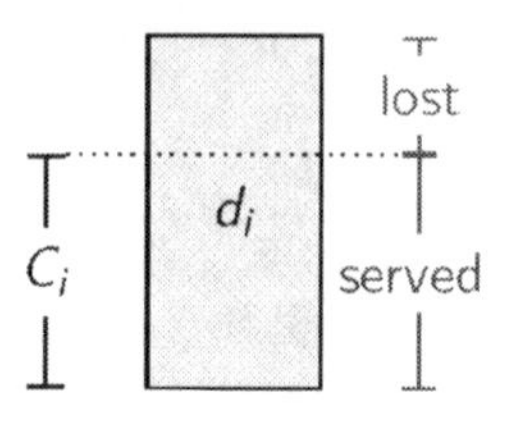

$$\mathbb{P}(\text{successful transmission}) = \min\left(1, \frac{C_i}{d_i}\right)$$

$$\Rightarrow \text{Expected number of transmissions} = \frac{1}{\mathbb{P}(\text{success})} = \max\left(1, \frac{d_i}{C_i}\right)$$

Fig. 3. Description of the loss and retransmission model

Users try to choose the providers with the cheapest perceived price. According to Wardrop's principle, all providers with positive demand have the same perceived price $P= P_i$, otherwise some users would have an interest to switch, and operators with a price P_i larger than P have no demand (they are too expensive). We assume that there is a total demand function D that depends on the perceived price P, and which is continuous, strictly decreasing on its support. Total demand is also equal to the total sum of received packets. It can be shown (see [10]) that there exists an equilibrium demand $d=(d_i)_i$ and that the perceived price is unique at that equilibrium.

Using this equilibrium on the demand distribution, providers play (by anticipation) their pricing game, trying to maximize their revenue R_i, knowing that increasing their price could incentivize users to move to competitors. The revenue R_i is the product of demand d_i and price p_i. Under assumptions on the elasticity of demand (of absolute value larger that 1), or if there are management costs that depend on the managed demand at each operator, we have been able to prove that there is a unique Nash equilibrium to the price war, described by the equations below. In other words, at equilibrium, all providers play the same price, and that price is such that demand equals capacity.

$$\forall i, \quad \begin{cases} p_i &= D^{-1}\left(\sum_j C_j\right) \\ d_i &= C_i. \end{cases}$$

3 Capacity/Technology Planning and Pricing in a Competitive Environment

Pricing is an issue for providers, but capacity planning is another, at a higher level: is it worth investing in capacity in a competitive network? This is even more true if congestion pricing is applied, i.e., if customers pay more when there is congestion; in that case it might be worthwhile to voluntarily have a congested network.

In [10], we have shown that under an assumption of sufficiently high demand elasticity, there is no interest for providers to lie on their capacity by declaring less than they actually have: they get higher revenues if using all their resource.

In [11,12] we have even played an additional game at a higher level, on top of the two-levels game we have just described. The idea is to look at the investment strategies of providers, in terms of acquiring new infrastructures and licenses, in new technologies for instance (WiMAX, new 3G license, etc.), and to maintain or expand the ones they might already own (WiFi, 3G...). Providers are usually heterogeneous because they already own different infrastructures and licenses and have different costs. We have thus defined a so-called third level technological game where each provider i chooses the subset S_i of implemented technologies (in a set defined as {WiFi, WiMAX, 3G}), resulting in a (multidimensional) matrix of revenues $(R_1(S), R_2(S))$ from the above game, for all combinations of (S_1, S_2), with $S=(S_i)_i$. We define similarly a cost matrix $(C_1(S), C_2(S))$ also for all combinations of (S_1, S_2) providing the cost for implementing the considered subsets of technologies. The goal of each provider i is to maximize his net benefit $B_i(S)= R_i(S)-C_i(S)$.

Several scenarios are presented to illustrate the best investment strategies of providers in games representing for instance an already positioned 3G against a rather WiFi-positioned one, and their relative strategies with respect to WiMAX/LTE. Another issue is at which maximal 3G license price a new entrant would be willing to implement this technology and develop the corresponding architecture. (Note that in [11,12] we have not used the perceived price we have described in Section 5.3.2 but a model without loss but delay at a queue and a perceived price sum of the price paid and a perceived delay cost. Though, this difference does not change the analysis in itself.

4 Analysis of the Price of Anarchy

It is well known from Game Theory that the conjunction of selfish decisions will not necessarily be globally optimal, and indeed can be worse for every participant of the game than some other possible outcome. However, by nature networks involve a very large number of actors of several kinds (consumers, providers, brokers), with diverging interests, so it is unreasonable to suggest that the network be centrally controlled to improve its efficiency. Therefore, the loss of efficiency due to selfish behaviors of actors cannot be ignored. This loss can be quantified by the so-called Price of Anarchy, recently defined [4] as the worst-case ratio comparing the global efficiency measure (that has to be chosen) at an outcome of the non-cooperative game played among actors, to the optimal value of that efficiency measure. If the price of anarchy is not too large, then this suggests that the system is run close to the optimum without the need for any coordination. Otherwise, some tools should be used to limit the efficiency loss (via the introduction of incentives in the network).

Remarkably, for the model presented in Section 2 (and most of the models presented in the next part), see [10], the price of Anarchy is 1 when trying to maximize the sum of utilities of all actors in the game, meaning that there is no loss of efficiency due to competition with respect to the case where all providers cooperate.

5 Analysis of Specific Other Competition Contexts

Considering the model described in Sections 2 and 3 again, but with a provider covering a smaller part of the population. It can model for instance a WiMAX (or LTE) operator against a WiFi operator as described in Figure 4. A proportion α of the population (i.e., of the demand function) is in the domain covered by the WiFi (thus by both operators) while a proportion 1-α is only covered by the WiMAX operator.

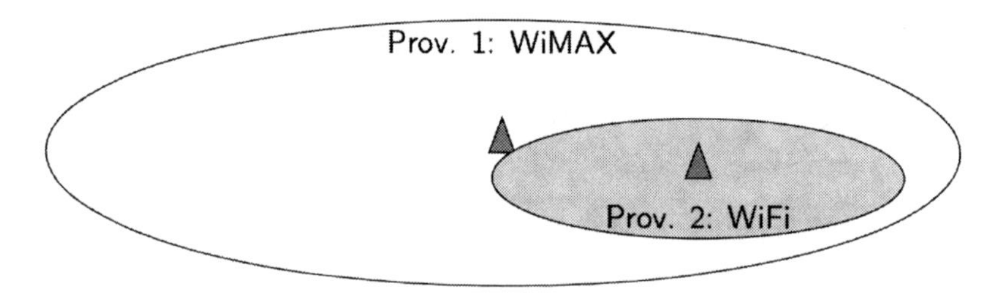

Fig. 4. Game between providers, one covering a smaller part of the population

Here again, the game is played as a two-levels game where the providers play on prices first, trying to maximize their revenue and anticipating the result of the game between users who distribute themselves between users trying to choose the provider with the cheapest perceived price, if not too expensive for them. Here also, it can be shown (see [8] for all details) that for all price profile, there exists at least a user (Wardrop) equilibrium. Moreover, the corresponding perceived prices of each provider are unique. At the level of price war between providers, using the user equilibrium to describe demand distribution, it is shown under conditions on demand elasticity similar to the previous subsection that there exists a unique Nash equilibrium (p_1^*, p_2^*) on prices:

- If the proportion α of population that the WiFi provider can reach is smaller than the proportion of capacity owned by this WiFi provider $C_2/(C_1+ C_2)$, then the common area is (entirely) left to the WiFi provider by the WiMAX one, which set a higher price and focus on the zone where he is in a monopoly. The equilibrium prices are

$$p_1^* = D^{-1}\left(\frac{C_1}{1-\alpha}\right) \geq p_2^* = D^{-1}\left(\frac{C_2}{\alpha}\right).$$

- If on the other hand the proportion α of population in the common zone is larger than the proportion of capacity owned by the WiFi provider, at equilibrium there is a price war between providers, which set the same price, and share the common area. This common price is given by

$$p_1^* = p_2^* = p^* = D^{-1}(C_1 + C_2).$$

Another extension, detailed in [7], studies the situation when each provider i has the capacity to serve C_i packets (or units) per slot, but when there is a capacity C that can be freely used by all providers. This basically helps to model the case when providers have their own licensed spectrum, but can use a free spectrum (for example WiFi) to send data when they are congested. Considering only two providers, the idea is that each of them uses his own capacity but sends demand in excess on the shared capacity C. If this capacity is overused then, there again, lost packets are chosen uniformly (which means for each provider proportionally to the amount he has sent to his shared spectrum). This is summarized in Figure 5, showing how the shared capacity is split between the two providers.

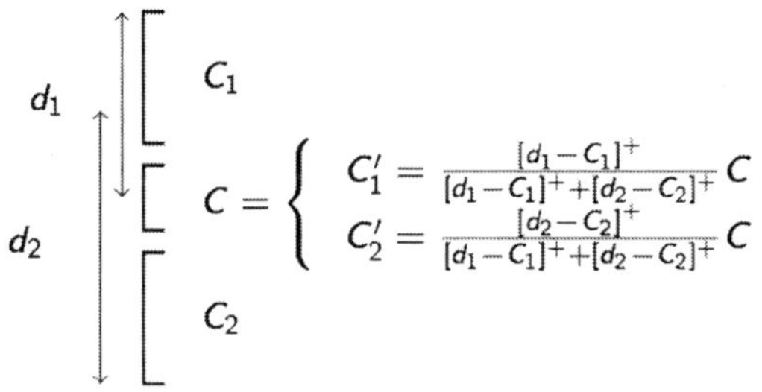

d_1 C_1

d_2 C_2

$$C = \begin{cases} C'_1 = \frac{[d_1 - C_1]^+}{[d_1 - C_1]^+ + [d_2 - C_2]^+} C \\ C'_2 = \frac{[d_2 - C_2]^+}{[d_1 - C_1]^+ + [d_2 - C_2]^+} C \end{cases}$$

Fig. 5. Network usage with two providers when there is a shared capacity

Here again, the existence of a user equilibrium is proved, and the pricing game between providers can be studied. An interesting question to answer becomes: is it of interest to license spectrum or not? To see this, we can plot in Figure 6, for some numerical values, the utilities of all actors (aggregated users, also called user welfare (UW), revenues of the two providers, and social welfare (SW), made of utilities of users plus revenues of providers) at equilibrium in terms of the proportion μ=C/C_{tot} of shared capacity, with C_{tot}=C+ C_1+ C_2.

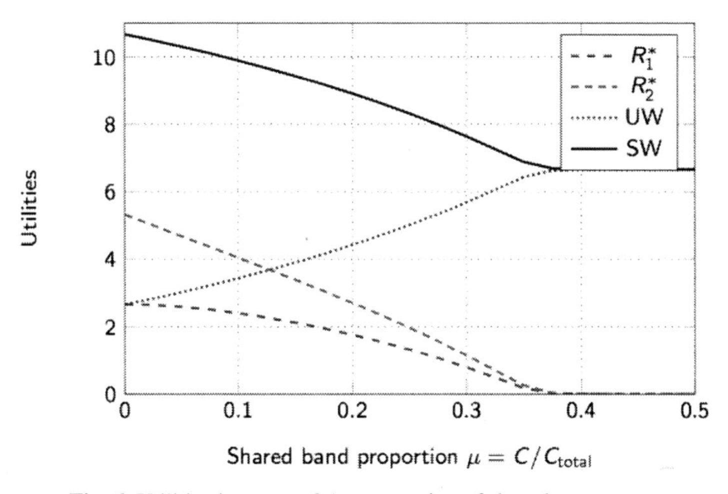

Fig. 6. Utilities in terms of the proportion of shared spectrum

What can be concluded from Figure 6 is that licenses are beneficial to providers (in our competitive environment) because their revenues decrease with μ, while it is the opposite for user welfare. Social welfare decreases too, but if we put a weight a bit more important on user welfare, an optimal intermediate situation can be obtained.

6 Other Competition Issues

Other competition issues can be treated and have been dealt with during Econ@Tel and CAPTURES projects. First, we have discussed in [5,6] competition in the case when customers churn, i.e., switch from one provider to another (this phenomenon has a considerable impact on network provider revenues due to its frequency). To prevent users from switching to another provider, operators can use retention strategies that have to be studied and sanctioned if they keep from an efficient use of the resources. We analyze the price war and regulation procedures (sanctions for providers) leading to the best equilibrium.

But competition is not only at the access provider level, it can be at the content provider level. The typical game is a content and advertising game (mainly how content providers earn money). As an illustration, we have studied in [9] two search engines (which could be Google, Yahoo! or Bing) in competition for advertisers (search engines usually displaying advertising slots when a search is performed), those advertisers having a fixed budget to split among search engines. The revenues of all actors depend on the market share of engines, their ranking and pricing rules, and the profile of strategies of advertisers. Here too, a two level game can be constructed with at the largest time scale the search engines choosing their ranking and pricing strategy, and at the lowest time scale, the advertisers splitting their advertising budget. Again the game is analyzed by backward induction, the engines being assumed to anticipate the reactions of advertisers. This helps to define the best ranking and pricing strategies *in a competitive context*, something not really studied in the area.

Other important issues are treated in other sections of this book, such as the competition for *security*, not really dealing with content or broadband/access but rather something transversal, and the *network neutrality* debate (discussed in a previous chapter), which aims at defining the relations between access providers and content providers, those entities trying to deviate from a cooperative behavior.

References

1. Courcoubetis, C., Weber, R.: Pricing Communication Networks|Economics, Technology and Modeling. Wiley (2003)
2. DaSilva, L.: Pricing of QoS-Enabled Networks: A Survey. IEEE Communications Surveys & Tutorials 3(2) (2000)

3. Greene, W.: Econometric Analysis. Prentice Hall, London (1997)
4. Koutsoupias, E., Papadimitriou, C.: Worst-Case Equilibria. In: Meinel, C., Tison, S. (eds.) STACS 1999. LNCS, vol. 1563, pp. 404–413. Springer, Heidelberg (1999)
5. Maillé, P., Naldi, M., Tuffin, B.: Competition for migrating customers: a game-theoretic analysis in a regulated regime. IEEE Globecom (December 2008)
6. Maillé, P., Naldi, M., Tuffin, B.: Price war with migrating customers. In: 17th Annual Meeting of the IEEE/ACM International Symposium on Modeling, Analysis and Simulation of Computer and Telecommunication Systems (MASCOTS 2009). IEEE Computer Society (September 2009)
7. Maillé, P., Tuffin, B.: Price war with partial spectrum sharing for competitive wireless service providers. IEEE Globecom (December 2009)
8. Maillé, P., Tuffin, B.: Price war in heterogeneous wireless networks. Computer Neworks 54(13), 2281–2292 (2010)
9. Maillé, P., Tuffin, B.: Sponsored search engines in competition: Advertisers behavior and engines optimal ranking strategies. In: 19th Annual Meeting of the IEEE/ACM International Symposium on Modeling, Analysis and Simulation of Computer and Telecommunication Systems (MASCOTS 2011). IEEE Computer Society (July 2011)
10. Maillé, P., Tuffin, B.: Competition among providers in loss networks. Annals of Operations Research (2011) (to appear)
11. Maillé, P., Tuffin, B., Vigne, J.: Economics of technological games among telecommunication service providers. IEEE Globecom (December 2010)
12. Maillé, P., Tuffin, B., Vigne, J.: Technological investment games among wire-less telecommunications service providers. International Journal of Network Management 21, 65–82 (2011)
13. Marbach, P., Berry, R.: Downlink resource allocation and pricing for wireless networks. In: Proc. of IEEE INFOCOM (2002)
14. Osborne, M., Rubenstein, A.: A Course on Game Theory. MIT Press (1994)
15. Tuffin, B.: Charging the Internet without bandwidth reservation: an overview and bibliography of mathematical approaches. Journal of Information Science and Engineering 19(5), 765–786 (2003)
16. Wardrop, J.: Some theoretical aspects of road traffic research. Proceedings of the Institute of Civil Engineers 1, 325–378 (1952)

Economic Traffic Management: Mechanisms and Applications

Fabio Hecht and Burkhard Stiller

University of Zürich, Communication Systems Group CSG, Switzerland
{hecht,stiller}@ifi.uzh.ch

Abstract. The paradigm shift from centralized services into hundreds of thousands of decentralized services changes traffic directions and patterns in the Internet. In that light, overlay providers form logical communication topologies, which effect the underlying telcos' as well as Internet Service Providers' (ISP) traffic and network management. Thus, Economic Traffic Management (ETM) principles deal with incentives, decentralized traffic generation, and modern applications, such as integrated Video-on-Demand and live streaming approaches. Therefore, this work overviews the very recent work on ETM mechanisms, especially driven by the SmoothIT project, and on LiveShift, combining efficiently architectural and system demands for an incentive-driven application management. Furthermore, this section describes in examples, provides evidences, and concludes that ETM and LiveShift can pave an efficient path to modern application support of distributed overlay traffic management, which benefits telcos, ISPs, overlay providers, and end users, if all or parts of them see the right incentives to cooperate, behave well, or interact.

Keywords: Economic Traffic Management, distributed overlay traffic management, incentives, P2P, video-on-demand, live streaming.

1 Introduction

The Internet traffic is still doubling approximately every 18-month; up to 80% of this traffic stems from Peer-to-Peer (P2P) and other overlay network-based applications [10]. Since bandwidth and reachability of end-nodes is also increasing, *e.g.*, due to the deployment of Fiber-to-the-Home (FTTH) access, overlay-based applications are expected to continue contributing to a large share of Internet traffic in the future. This indicates that a paradigm shift from services offered centrally to services offered by end-nodes is already happening, and most network traffic will follow network paths forming logical network overlays at the service layer, such as Virtual Private Networks, Network Management System overlays, or distributed collaboration overlays. For today's Telecommunication Service Providers (telco) and Internet Service Providers (ISP) the crucial issue of how to control and manage network traffic stemming from these overlay-based applications arises.

The use of Economic Traffic Management (ETM) mechanisms based on incentives for controlling and managing network traffic of overlay applications in the Internet has

A.M. Hadjiantonis and B. Stiller (Eds.): Telecommunication Economics, LNCS 7216, pp. 188–198, 2012.

been researched very recently and is still in its early stages [21]. However, initial results have shown that such mechanisms do have the important property of scalability with respect to the number of players (*i.e.* providers, consumers, or peers), and also that they do lead to a more efficient network operation. In managing traffic created and routed through their networks, today's ISPs, which may be service providers on the basis of IP (Internet Protocol), employ methodologies suitable for traditional traffic/service profiles. As the structure of overlays determines the traffic flows in ISP networks, it is highly desirable for ISPs to influence the overlay configuration based on information on their specific structure or use. Thus, overlays must be managed to maximize the advantages for multiple ISPs involved (independently of underlying topology), increase their capability to withstand faults, and balance network load.

From a user's perspective, ETM mechanisms show the potential to be applied to a wide range of services [21]. Those services may differ with respect to Quality-of-Service (QoS) metrics and may be charged differently. This enables a market of competitive overlay services and their management. Users will benefit from this competition and the variety of services and service qualities, made conveniently available to users.

SmoothIT [20] has investigated several ETM mechanisms to bridge the gap between overlay and underlay to benefit ISPs, overlay providers, and overlay users [21]. ETM provides the relevant decentralized control technology and management approaches, backed by thorough theoretical investigations, which enable new service quality levels. The application of such scalable ETM mechanisms enables ISPs to reduce their service provisioning and maintenance costs. At the same time, since other overlay applications often cover end-users connected to multiple ISPs, appropriate incentives schemes for collaboration among ISPs (which may also be competing ones at the same time) have been defined and investigated thoroughly. In addition, the incentives need to match detailed application requirements, to be applicable and successful.

Thus, the LiveShift approach determines a P2P application [11] forming an overlay on its own — supporting both live and time-shifted steaming in an integrated manner. It allows for the collaboration beyond the current state of the art, enabling peers to store video streams received in order to distribute them in the future. A highly relevant use case of this functionality allows users, without having previously prepared any local recording, to watch programs from the start or to jump over uninteresting parts until seamlessly catching up with the live stream. Watching pre-recorded or live content does not require different protocols and system architectures — LiveShift provides these two functionalities —, however, up to now only two incompatible incentive schemes exist, which cannot cope with this LiveShift integration of live and time-shifted streams. Thus, LiveShift provides an interesting use case for the deployment of the ETM mechanisms defined, since the feature of storing and a later distributing video streams increases the opportunities of peer collaborations.

The remainder of this section presents in Sect. 2. major related work, which is combined with a discussion of incentives in distributed systems. In support of the feasibility of ETM and an integrated LiveShift streaming system, the two architectures of the system and applications side are defined in Sect. 3.. These development steps are complemented by a selected set of results obtained throughout the respective evaluation in Sect. 4. Finally, Sect. 5. summarizes the key findings and draws conclusions.

2 Related Work and Incentives in Distributed Systems

Related work in the field of ETM and incentives has addressed only very recently this research domain of integrating economically-driven approaches with very specialized optimization tasks. Thus, key approaches are sketched and briefly discussed.

Thus, [13] described early concepts that can be considered to be a very early and limited instance of ETM, which is applied to overlay traffic. Furthermore, the Knowledge Plane [6] defines a system for gathering observations, constraints, and assertions, which applies rules to these to generate observations and responses for network operation. As such it provides a basis for the autonomy in ETM. More specifically, the concept of Biased Neighbor Selection [3] introduces the idea of adapting overlays to explore traffic localization. The Ono plug-in approach [5] implements this concept as part of a P2P file-sharing application, exploiting information obtained indirectly from Content Distribution Network (CDN) servers. In addition, a few approaches have been proposed, which try to facilitate the collaboration between P2P applications and ISPs. The P4P project [22] proposes the presence of an information server, called iTracker, to provide locality information to the overlay. The iTracker may communicate directly with peers or application trackers. A similar approach, the Oracle service [2], runs at the ISP and influences the overlay by ordering potential neighbors by the ISP's preference. The ALTO (Application-Layer Traffic Optimization) IETF Working Group [1] is working on defining a standard for such services provided by ISPs to overlay applications. The area of incentives in distributed systems, particularly for P2P systems, has attracted significant attention in the past. Many approaches are interested in providing incentives for peers to share their resources [19], *e.g.*, upload files or provide content. Incentive schemes, such as Tit-for-Tat [7] and CompactPSH [4], rely on reciprocity, which is challenging in a video straeming system, since the stream is bound to flows within a single direction only (in most cases). A different approach is relying on reports sent by peers [14], possibly having a way to verify those reports [16]. The problem is challenging in LiveShift, since it supports live streaming, in which rate-based incentives are appropriate, and fully integrated video-on-demand, in which peers need to store streams to be provided, thus relying on immediate reciprocation is unsuitable. ETMs go beyond this definition to provide incentives for multiple players to support the distributed service provision. In that sense the alternative choices for ETMs are outlined in further detail in Sect. 3.1 below, since they have been developed in theory and implementation by the SmoothIT project.

3 ETM and Application Architecture

To reach a suitable approach for meeting the set of requirements as outlined, (a) relevant ETM mechanisms developed during the SmoothIT project and (b) a description of the architectures of the ETM System and LiveShift, as a possible application for the deployed ETM mechanisms, has to be defined. In that respect these areas are outlined below to provide the basis for a subsequent evaluation.

3.1 ETM Mechanisms

To enable overlay traffic management techniques, which include economically-driven incentives, relevant stakeholders' (users, ISPs, and overlay providers) and their evaluations are required to be determined. In that sense so-called "win"-situations define that a stakeholder does see a positive effect upon applying an ETM mechanism. In the best cases, the application of ETM mechanisms lead to a TripleWin situation, in which all three stakeholders are identified to benefit from ETM. The most important of these can be classified into the following three main categories [21]:

Locality promotion involves all peers of a domain, in which they rate overlay neighbors according to their underlay proximity. Within the BGPLoc (border Gateway Protocol Locality) approach a peer sends a list of peers to the SIS (SmoothIT Information Service) [18]. The latter interfaces with the underlay network, obtains locality information, and rates each peer based on certain location attributes, *e.g.*, on the number of hops in the Autonomous System (AS) path. This rated and accordingly sorted list is returned to the requesting peer where dedicated overlay operations (*e.g.*, unchoking and neighbor selection in BitTorrent) are modified to favor peers belonging to the same domain with the requesting one. Thus, interconnection costs for ISPs are reduced, while the performance for peers is also improved in a variety of cases.

With the **insertion of locality-promoting peers** or resources an indirect way of promoting locality for the ISP is introduced [15] . This is based on a domain-special resource, such that content is downloaded faster into the domain and distributed among local peers. Thus, in IoP-based (ISP-Owned Peer) solutions the IoP initially participates in the swarm as a normal peer, but quickly acquires the entire content and starts serving the local peers. Two instances of such ETM mechanisms have been developed, besides the IoP the Highly Active Peer (HAP). While in case of the IoP the ISP deploys and controls a peer with augmented bandwidth and storage capabilities (allowing for a download of content and making it available to domain's peers), the HAP this idea is implemented in a decentralized manner, with the functionality being passed to normal peers. More specifically, due to overlay and underlay information, certain highly active peers in a domain are promoted by the ISP to so-called enhanced peers and their Internet connection speed is increased dynamically by the ISP, mainly to enable a faster download and to offer overlay content efficiently.

The **inter-domain collaboration** (Inter-SIS) addresses cases, where information available to a domain is not sufficient to reach some of the aforementioned optimization objectives [8], [9]. This can be due to the asymmetry of routing information between source and destination domains or it is due to the lack of connection details of communicating peers. In such cases, inter-domain collaboration can lead to fine-tuning and improvement of local decisions in all participating domains.

3.2 SmoothIT's ETM System Architecture

To provide for a suitable, seamlessly integrated, and efficient architecture for ETM mechanisms in today's and tomorrows Internet Fig. 1 shows the relevant architecture developed to support the overlay and underlay integration through the so-called ETMS (ETM System) [21]. Within the ETMS layer, the SIS (SmoothIT Information Service) typically runs in ISPs premises.

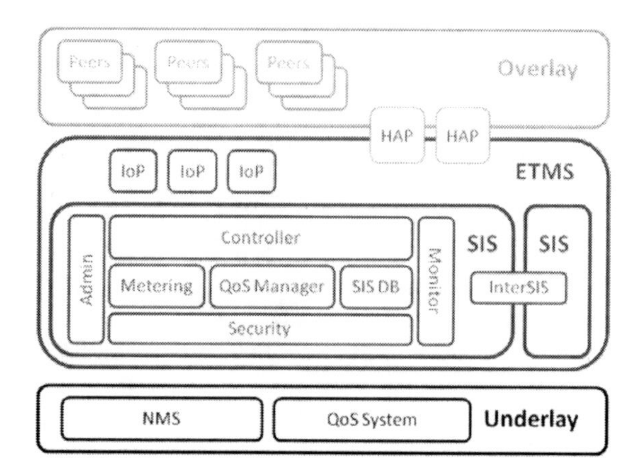

Fig. 1. SmoothIT ETM System Architecture

The Controller component determines the core of the system: it is responsible for the coordination of the ETM mechanism by receiving requests from overlay applications, the calculations based on the ETM mechanism chosen, and in cooperation with several underlay metrics (*e.g.*, provided by metering and policy information) for the output of the required information to the overlay application. This information returned is in form of ranked lists, such as rating peers as possibly "good" neighbors or as rating swarms as "popular" ones.

The Metering component is responsible for collecting network information from the underlying Network Management System (NMS) and supplying them to the ETM mechanisms in the SIS. This information can include, *e.g.*, BGP (Border Gateway Protocol) routing information, network performance parameters, and user profiles. Furthermore, the QoS Manager checks the availability of network resources, and may be used also to guarantee resources requested by the end user, enforcing QoS policies in the network. It interfaces to the network by using, for example, NGN (Next Generation Network) transport control functionalities. Additionally, the InterSIS component facilitates the exchange of information between two or more SISs to fine-tune local decisions. Finally, the SIS DB (Database) stores all information that may be useful for all modules, such as configuration parameters or business logic. Further details can be obtained from [21].

3.3 LiveShift's Application Architecture

LiveShift is introduced as a promising example application to fit the ETM architecture as introduced at the overlay layer, since it provides both highly dynamic content (the live stream) and content that can be stored (the distributed storage of past live streams). Using the ETMS on LiveShift does show a great potential, due to the inherent difficulty in providing the right incentives, mainly due to the mixed scenarios of high traffic loads generated and the delay-sensitivity present in video streaming, all of which influencing the quality-of-experience of the end user. LiveShift follows the mesh-pull approach, in which peers split video streams into chunks, and exchange them with no fixed structure, using (1) a publish/subscribe protocol to report chunk availability and (2) a distributed tracker for content and peer discovery [11].

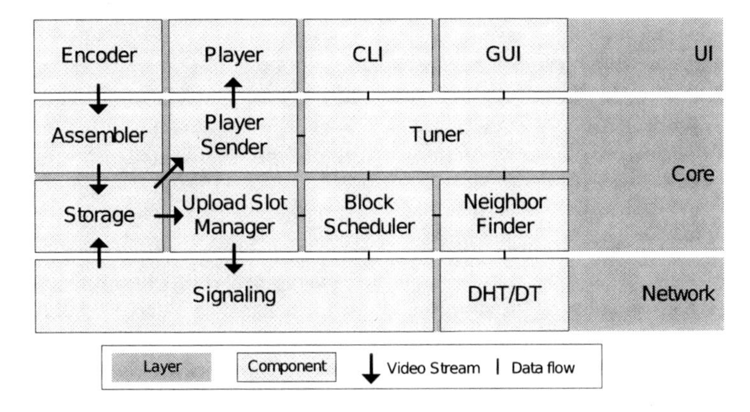

Fig. 2. LiveShift architecture

While Fig. 2 shows major components of the LiveShift architecture, which are split into three layers: the Network layer (which corresponds to the Overlay in Fig. 1) contains the Distributed Tracker (DT) and the Distributed Hash Table (DHT) to store network-wide persistent information in a fully-decentralized manner, and a signaling component to handle direct communication between peers. The DT and DHT are provided by the TomP2P library [2]. The signaling component handles direct communication between peers. Peers use it to send unicast messages to other peers.

The main challenge in LiveShift (and in fact any P2P video streaming system) is to locate and download streams on time to keep up with the playback and with the added challenge of a potentially large time-scale. While incentive mechanisms in their original form (application-centered only) show limitations in providing both rate-based and long-term incentives for peers to share, ETM mechanisms do help in this respect by avoiding congestion, suggesting peers with lower latency or higher available bandwidth, caching content for time-shifting, and further providing incentives for peers to share. Thus, a combination of ETM and LiveShift shows valuable benefits.

4 Evaluation and Results

Besides the design of this emerging and innovative functionality, the evaluation and suitability needs to be proven, too. Thus, a selection of major results obtained with the introduced ETMS is outlined in Sect. 4.1, focusing on results on the BGPLoc ETM mechanism. And to complete, Sect. 4.2 contains selected LiveShift results, which address the proposed use case from above and which have been derived from an implemented prototype.

4.1 ETM Samples

The ETM system designed has been implemented prototypically [2]. The overlay application used there is based on Tribler, a P2P file-sharing application that implements the BitTorrent protocol. The modified application includes the BGPLoc ETM mechanism, which supports locality promotion at two peer selection algorithms: Biased Neighbor Selection (BNS) prioritizes selecting local peers for downloading, Biased Optimistic Unchoking (BOU) peers prefer local peers to grant their optimistic unchoking slot. For more information on the BitTorrent protocol, refer to [7].

The following results were obtained by simulating one BitTorrent swarm for the distribution of a file with size 154.6 MB for 6.5 hours [18]. The simulated topology consists of 20 Tier 3 stub ASes, two Tier 2 hub ASes, and one Tier 1 transit AS. Each stub, which AS runs an instance of the ETMS and a number of peers, and is directly connected to two other stub-ASes (depicting peering links) and one hub-AS. The two hub-ASes are interconnected via the transit-AS (transit links). Peers join the swarm at a random (uniformly distributed) stub AS with an exponentially distributed inter-arrival time with a mean value of 10s, and remain online until they have downloaded the entire file plus an additional, exponentially distributed seeding time with a mean value of 10 minutes. In all experiments, four different peer behaviors are compared: regular BitTorrent (regBT), BitTorrent with Biased Unchoking (BOU), Biased Neighbor Selection (BNS), and both (BNS&BOU) [18].

Additionally, the application supports live monitoring by periodically sending reports to the Monitor component, including overlay statistics (*e.g.*, total download and upload, download and upload rate, or download progress), and also video playback statistics (*e.g.*, play and stall time), which are stored in the SIS Database. All components of the SIS have been implemented in Java and are deployed on a JBoss application server. The current version of the Controller supports the BGP-based locality promotion (BGPLoc) and IoP mechanisms.

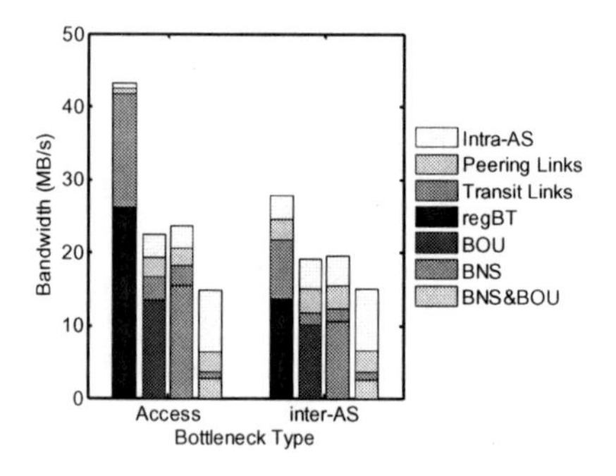

Fig. 3. Mean BitTorrent traffic (ISP) [18]

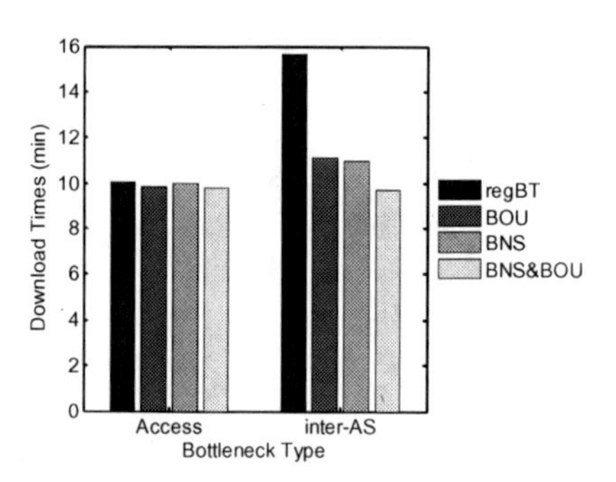

Fig. 4. Mean download time (end user) [18]

Fig. 3 and Fig. 4 consider that the networks suffers from a bottleneck at two different points: while the "Access" bottleneck assumes that there is no congestion in peering and transit links, the "Inter-AS" bottleneck considers that traffic shaping takes place between stub and hub ASs. Fig. 3 shows the traffic measured at different link types: the ISP is always able to save costs by reducing traffic at transit links, since intra-AS connections are preferred. Also, these savings are larger with the combination of BNS&BOU. Fig. 4 shows that download times remain unaffected, if the bottleneck is at the access link, but experiences a worse performance with regBT and inter-AS bottleneck. For results obtained with other ETM mechanisms, especially IoP, Inter-SIS, and HAP, refer to [15], [8], [9], and [17], respectively.

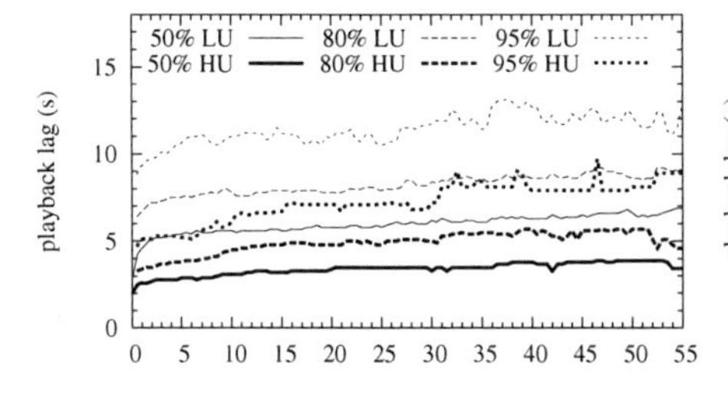

Fig. 5. Experienced playback lag vs. holding time in scenario s1

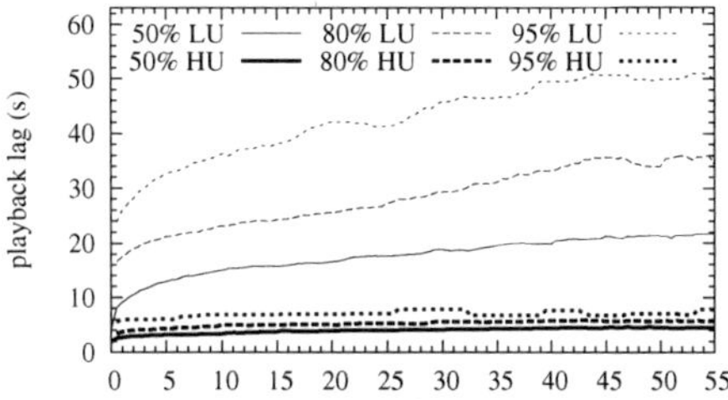

Fig. 6. Experienced playback lag vs. holding time in scenario s3

4.2 LiveShift Samples

LiveShift has been fully implemented in Java 1.6 and evaluated in a test-bed using traces to model user behavior. A complete definition of test-bed, scenarios, and user behavior can be found at [11]. A number of users were emulated, switching among 6 channels, and reporting metrics such as their playback lag or channel holding times.

Fig. 5 shows the experienced playback lag for scenario s1, which is composed out of 6 peercasters (peers that are sources for a channel, can upload 5 times the video stream), 15 high upstream peers (HU, can upload 5 times the video stream), and 60 low upstream peers (LU, can upload 0.5 times the video stream). Scenario s1, being well-provisioned with bandwidth, shows a low (i.e. less than 15s) playback lag for sessions of up to one hour for 95% of all HU and LU peers. Fig. 6 outlines values for Scenario s3, which, in contrast with s1, has 120 LU peers, thus, being on the edge of being underprovisioned in terms of upload capacity. Although the playback lag is considerably higher than on s1, which is natural, considering that it will take longer for peers to find available bandwidth, and the average hop count that every chunk needs to travel is higher, on average. The playback lag in s3 remains below 60 s for 95% of all peers.

Fig. 7 shows that (a) the overhead introduced by LiveShift is small (less than 3%), since the video stream it transports is already very bandwidth-consuming, and (b) that in Scenario s3 the share of failed playbacks (that is, the user received less than half of all video blocks in a moving window of 30 s and gave up) is already larger than 4%.

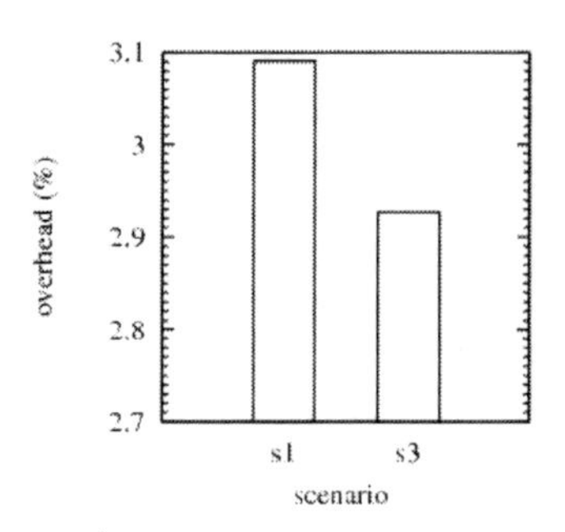

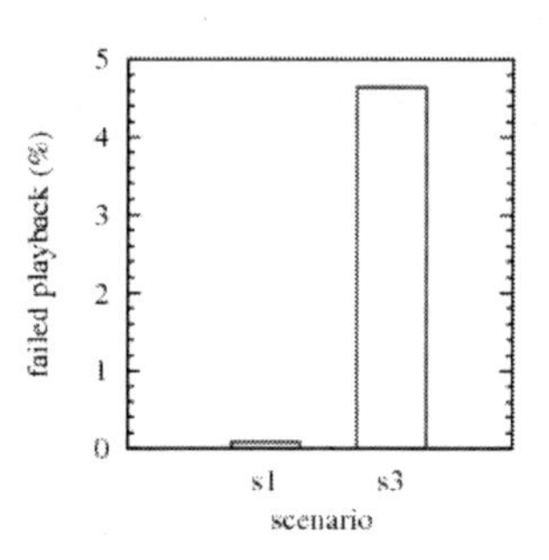

Fig. 7. Overall overhead and average failed playback in scenarios s1 and s3

5. Summary and Conclusions

This section describes a set of ETM (Economic Traffic Management) mechanisms defining alternatives to promote (directly or indirectly) localization of overlay traffic, namely the BGP-Loc, the IoP, and the Inter-SIS mechanisms. In combination with the LiveShift application, such modern overlays and respective functionality can support a scalable and distributed traffic and application management in the Internet.

To enlighten the feasibility of these approaches introduced, designed, and discussed, the respective architectures of the ETM System and LiveShift have been deployed in terms of a practical implementation performed in the SmoothIT and LiveShift approaches. Also, the key concepts, the use case, and evaluation results as presented show their potential in profiting from the presented ETM mechanisms and related incentives. In that sense, these approaches are backed by selected results from test-bed trials and performed evaluations.

Concluding, the application of economically-driven incentive mechanisms in a distributed networking environment show besides their implementation feasibility the advantages for different stakeholders under different overlay set-ups. In this variety of evaluations the promotion of locality information is highly beneficial, while avoiding the violation of any stakeholders' own optimization (TripleWin). Especially and in case of video sharing via BitTorrent, the combination of different peer behaviors show winning mechanisms, which is reachable by an flexible architectural support. Additionally, the behavior of the streaming/Video-on-Demand approach LiveShift indicates that acceptable playback delay are achievable, which does lead to valuable user experience. This is complemented by the fact that also for LiveShift the overhead introduced is very small, *e.g.*, in terms of failed playback delays. In turn, the (a) feasibility of those approaches, (b) the technical integration seamlessly into existing networking infrastructures, and (c) the achieved technical performance indicate that the ETM key idea combined with the incentives aspect show a beneficial outcome for overlay management principles.

Acknowledgement. This work has been performed partially in the framework of the EU Project SmoothIT (FP7-2007-ICT-216259), the COST Action IS0605, the EU Project SESERV (FP7-2010-ICT-258138), and the project LiveShift at CSG@IFI. The authors like to acknowledge particularly T. Hoßfeld, S. Oechsner, F. Lehrieder, K. Pussep, M. Michel, S. Soursos, I. Papafili, G. Stamoulis for providing partial results, and all members of SmoothIT and SESERV for their collaboration and support.

References

1. Application-layer traffic optimization (alto), IETF Working Group (2009), `http://www.ietf.org/html.charters/alto-charter.html`
2. Aggarwal, V., Feldmann, A., Scheideler, C.: Can ISPs and P2P systems co-operate for improved performance? ACM Computer Communication Review 37(3), 29–40 (2007)
3. Bindal, R., Cao, P., Chan, W., Medval, J., Suwala, G., Bates, T., Zhang, A.: Improving Traffic Locality in BitTorrent via Biased Neighbor Selection. In: 26th IEEE International Conference on Distributed Computing Systems, Lisbon, Portugal, p. 66 (July 2006)
4. Bocek, T., Hecht, F.V., Hausheer, D., Stiller, B., El-khatib, Y.: CompactPSH: An efficient transitive TFT incentive scheme for Peer-to-Peer Networks. In: 34th Annual IEEE Conference on Local Computer Networks (LCN 2009), Zürich, Switzerland, pp. 483–490 (October 2009)
5. Choffnes, D.R., Bustamante, F.E.: Taming the Torrent: A Practical Approach to Reducing Cross-ISP Traffic in Peer-to-peer Systems. ACM Computer Communication Review 38(4), 363–374 (2008)
6. Clark, D., Partridge, C., Ramming, J., Wroclawski, J.: A Knowledge Plane for the Internet. ACM Computer Communication Review 33(4) (October 2003); ACM SIGCOMM, Karlsruhe, Germany, pp. 3—10
7. Cohen, B.: Incentives Build Robustness in BitTorrent. In: 1st Workshop Economics of Peer-to-Peer Systems (P2PCON), Berkeley, California, U.S.A. (June 2003)
8. Dulinski, Z., Stankiewicz, R., Wydrych, P., Cholda, P., Stiller, B.: Inter-ALTO Communication Protocol. IETF Internet-Draft, draft-dulinski-alto-inter-alto-protocol-00 (June 2010); work in progress
9. Dulinski, Z., Wydrych, P., Stankiewicz, R.: Inter-ALTO Communication Problem Statement. IETF Internet-Draft draft-dulinski-alto-inter-problem-statement-01 (July 2011); work in progress
10. Haßlinger, G.: ISP Platforms Under a Heavy Peer-to-Peer Workload. In: Steinmetz, R., Wehrle, K. (eds.) P2P Systems and Applications. LNCS, vol. 3485, pp. 369–381. Springer, Heidelberg (2005)
11. Hecht, F.V., Bocek, T., Clegg, R.G., Landa, R., Hausheer, D., Stiller, B.: LiveShift: Mesh-Pull Live and Time-Shifted P2P Video Streaming. In: 36th IEEE Annual Conference on Local Computer Networks (LCN 2011), Bonn, Germany, pp. 319–327 (October 2011)
12. Lehrieder, F., Oechsner, S., Hoßfeld, T., Despotovic, Z., Kellerer, W., Michel, M.: Can P2P-Users Benefit from Locality-Awareness? In: 10th IEEE International Conference on Peer-to-Peer Computing 2010 (P2P 2010), Delft, The Netherlands (August 2010)
13. Liu, Y., Zhang, H., Gong, W., Towsley, D.: On the Interaction between Overlay Routing and Underlay Routing. In: INFOCOM 2005, Miami, Florida, U.S.A., pp. 2543–2553 (2005)
14. Mol, J.J.D., Pouwelse, J.A., Meulpolder, M., Epema, D.H.J., Sips, H.J.: Give-to-get: Free-riding-resilient video-on-demand in p2p systems. Multimedia Computing and Networking 6818, 681804—681804-8 (2008)

15. Papafili, I., Stamoulis, G.D.: A Markov Model for the Evaluation of Cache Insertion on Peer-to-Peer Performance. In: EuroNF NGI Conference, Paris, France (June 2010)
16. Piatek, M., Krishnamurthy, A., Venkataramani, A., Yang, R., Zhang, D., Jaffe, A.: Contracts: Practical Contribution Incentives for P2P Live Streaming. In: 7th USENIX Conference on Networked Systems Design and Implementation, Berkeley, California, U.S.A., September 2010, p. 6 (2010)
17. Pussep, K.: Peer-Assisted Video-on-Demand: Cost Reduction and Performance Enhancement for Users, Overlay Providers, and Network Operators; Ph.D. Thesis, Technische Universität Darmstadt, Germany (2011)
18. Racz, P., Oechsner, S., Lehrieder, F.: BGP-Based Locality Promotion for P2P Applications. In: 19th International Conference on Computer Communications and Networks (ICCCN 2010), Zürich, Switzerland, pp. 1–8 (August 2010)
19. SESERV Project (October 2011), `http://www.seserv.org`
20. SmoothIT Project (March 2011), `http://www.smoothit.org`
21. Soursos, S., Callejo Rodriguez, M., Pussep, K., Racz, P., Spirou, S., Stamoulis, G., Stiller, B.: ETMS: A System for Economic Management of Overlay Traffic. In: Tselentis, G., et al. (eds.) Towards the Future Internet — Emerging Trends from European Research, pp. 1–10. IOS Press, Amsterdam (2010)
22. Xie, H., Yang, R.Y., Krishnamurthy, A., Liu, Y.G., Silberschatz, A.: P4P: Provider Portal for Applications. ACM Computer Communication Review 38(4), 351–362 (2008)

Autonomic Management of Mobile and Wireless Networks

Antonis M. Hadjiantonis

KIOS Research Center for Intelligent Systems and Networks, University of Cyprus
antonish@ucy.ac.cy

Abstract. Autonomic management is receiving intense interest from academia and industry, aiming to simplify and automate network management operations. Autonomic management or self-management capabilities aim to vanish inside devices, relieving both managers and users from tedious configuration and troubleshooting procedures. Ideally, self-managed devices integrate self-configuration, self-optimization, self-protection and self-healing capabilities. When combined, these capabilities can lead to adaptive and ultimately self-maintained autonomic systems.

Keywords: autonomics, policy-based, self-management, network management.

1 Introduction

Mobile and wireless networks have become a ubiquitous reality and evermore surround our everyday activities. They form and disappear spontaneously around us and have become new means for productivity and social interaction. Access to corporate networks, e-mail or simply entertainment, these are the new necessities posed on an increasingly networked wireless world. In the era of mobility and connectivity, a multitude of devices interact with us in our everyday life. Wireless digital assistants such as mobile phones, laptops or personal organizers must be able to cope and offer the desired services at any place and at anytime. An increasingly ad hoc element facilitates the need for on demand connectivity and wireless communication. At the same time, increased complexity and heterogeneity have become barriers to seamless integration and ease of use.

In reality though, the deployment of self-managed networks is withheld from several obstacles that need to be overcome in order to realize such a vision. The use of policies for network and systems management is viewed as a promising paradigm to facilitate self-management. Policies can capture the high-level management objectives and can be automatically enforced to devices, simplifying and automating compound and time-consuming management tasks.

Nowadays, services targeting home and business users, such as Internet access, digital television or online entertainment, are taken for granted. However, modern lifestyle creates the need for extending the reach of such services but also creates the need for new ones, targeted to people on the move. The convergence of fixed and

A.M. Hadjiantonis and B. Stiller (Eds.): Telecommunication Economics, LNCS 7216, pp. 199–208, 2012.

wireless technologies is inevitable and a rapidly evolving market brings new challenges. In recent years, we have experienced an unprecedented penetration of mobile phones, while the mobile industry growth and evolution continues steadily. Developed countries are planning their transition to fully converged networks and services, while wireless access capability (Wi-Fi) is becoming a common feature of mobile phones. At the same time, newly industrialized countries and emerging markets are just discovering wireless technologies and offer an impressive drive for low cost infrastructure development. Their massive potential customer base is contrasted to low population density, making the cost of wired technologies prohibitive and deeming current management paradigms as inapplicable. At the same time, new wireless networking and management paradigms are investigated, offering a promising and challenging ground for research and innovation.

Improved network organization can increase scalability and decentralize management responsibilities, but one has to consider that the majority of wireless networked devices are not under the strict control of a network operator as in traditional infrastructure-based networks. Therefore, a critical management requirement is to respect the owner relationship between end-users and managed devices. Individual users are reluctant to entrust the command of their devices to an operator and demand more control. The lack of a single administrative authority complicates management tasks, but at the same time motivates research on collaborative management schemes. Open standards and contractual agreements can facilitate the interests of different managing entities, e.g. network operators or service providers. The goal is to provide an adaptive framework for network and service management, where users' privacy and preferences are respected, while multiple managing entities can offer services tailored to the users' needs.

The policy-based management (PBM) paradigm can provide the means to integrate self-management capabilities and policies can capture the high-level management objectives to be autonomously enforced to devices. Although the PBM paradigm has been traditionally employed in large-scale fixed IP networks, its controlled programmability can significantly benefit the highly dynamic environment of mobile and wireless networks. PBM can offer a balanced solution between the strict hard-wired management logic of current management frameworks and the unrestricted migration of mobile code offered from mobile agents. This has motivated the adoption of PBM for the autonomic management of mobile and wireless networks, aiming to simplify and automate compound time-consuming management tasks. The centralized orientation of policy-based operations requires significant research efforts to accommodate the needs of distributed policy-based management (D-PBM) for the next generation of mobile and wireless networks. In addition, in a rapidly evolving multi-player environment, policies can express the interests of different players and facilitate their cooperation. PBM can be a future-proof solution and can provide the flexibility to adapt to change. At the same time, the users' requirements for control and privacy can be encapsulated in policies and with minimum intervention their devices can operate autonomously.

2 Policy-Based Management (PBM)

Policy-Based Management (PBM) [2][15][13] and policies have been envisioned as encapsulating business objectives which in turn are autonomously applied to managed systems, requiring minimal human intervention. However, practice has shown that what was initially conceived as the instant panacea of network management is in fact a long journey towards self-managing networks, hampered by severe obstacles. The views published in [4] by a major infrastructure vendor are illustrative of initially overestimated expectations from policies: "to many people, it suggests that, by some magic, you get something for nothing, or at least without needing to think through what needs to be precisely done to achieve those objectives. Of course, there is no magic, and anyone expecting magic is bound to be disappointed". Beyond initially high expectations, research on PBM has gradually verified its enormous potential and showed that it can simplify complex management tasks of large-scale systems. The concept of high-level policies monitoring the network and automatically enforcing appropriate actions has received intense interest and has been fuelled by the renewed interest in Self-Management and Autonomic Networking [2][15][13].

In general, policies can be defined as Event-Condition-Action (ECA) clauses, where on event(s) E, if condition(s) C is true, then action(s) A is executed. Different definitions and classification of policies can also be found in the literature and are presented later. The main advantage which makes a policy-based system attractive is the functionality to add controlled programmability to the managed system, without compromising its overall security and integrity [5]. Real time adaptability of the system can be mostly automated and simplified by the introduction of the PBM paradigm. According to [5] policies can be viewed as the means to extend the functionality of a system dynamically and in real time in combination with its pre-existing hard-wired management logic. Policies offer the unique functionality to the management system of being re-programmable and adaptable, based on the supported general policy types. Policies can be introduced to the system and parameterized in real time, based on management goals and contextual information. Policy decisions prescribe appropriate actions on the fly, to realize and enforce those goals.

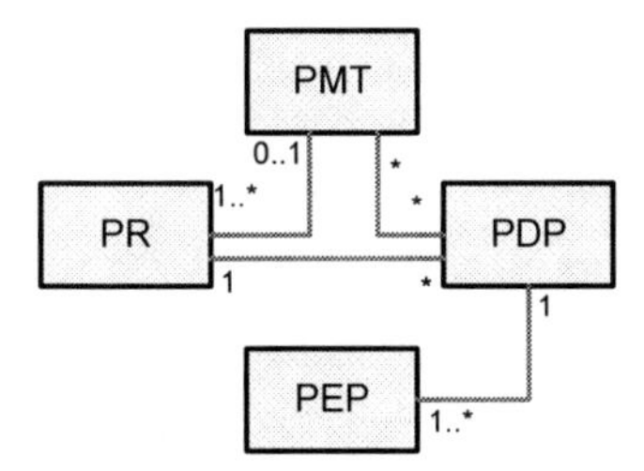

Fig. 1. PBM functional elements

A block diagram of PBM functional elements is shown in Figure 1, using a simplified UML notation of their relationships. These four elements constitute IETF's policy-based framework, as proposed through the work of the Policy Framework WG (POLICY) and the Resource Allocation Protocol WG (RAP):

- Policy Management Tool (PMT): the interface between the human manager (e.g. a consultant or network administrator) and the underlying PBM system.
- Policy Repository (PR): the blueprint of policies that a PBM system is complying with at any given moment. It encapsulates operational parameters of the network and therefore it is one of the most critical elements.
- Policy Decision Point (PDP): a logical entity that makes policy decisions for itself or for other network elements requesting such decisions. These decisions involve on one hand evaluations of policy rule conditions and on the other hand deal with the actions' enforcement when conditions are met.
- Policy Enforcement Point (PEP): a logical entity that enforces policy decisions. Traditionally, the sole task of PEP is to execute policy decisions, as instructed by the controlling PDP.

The IETF framework is widely used and accepted in research and industry and has served as a reference model for PBM systems [16],[2]. Managing Entities use a Policy Management Tool (PMT) to introduce and store policies in the Policy Repository (PR). The PR is a vital part for every policy-based system because it encapsulates the management logic to be enforced on all networked entities. Stored policies can be subsequently retrieved, either by Policy Decision Points (PDP) or by another PMT. Once relevant policies have been retrieved by a PDP, they are interpreted and the PDP in turn provisions any decisions or actions to the controlled Policy Enforcement Points (PEP).

Policy provisioning is the process of communicating policy decisions and directives between a Policy Decision Point (PDP) and a Policy Execution Point (PEP) using a suitable protocol (RFC2753, RFC3198). A PDP is also known as a policy server, reflecting its responsibility to serve a number of PEP with policy decisions and relevant PBM information. On the other hand, PEP are also known as policy clients since their operation depends on these decisions, as provided by their parent PDP. The protocol involved in this communication is the policy provisioning protocol. Efforts from IETF's Resource Allocation Protocol Working Group (RAP WG) have produced the COPS (Common Open Policy Service) Protocol (RFC2748) and COPS protocol for Policy Provisioning (COPS-PR) (RFC3084). COPS is a simple query and response protocol that can be used to exchange policy information between a policy server (PDP) and its clients (PEP). The basic model of interaction between a policy server and its clients is compatible with IETF's policy-based framework. The focus of IETF's efforts has been mainly to provide a protocol to carry out the task of policy provisioning mostly related to QoS parameters and setup. Beyond COPS, no other dedicated policy provisioning protocol has been standardized by the IETF and policy provisioning has been viewed under the general umbrella of configuration management protocols. Traditional management protocols (SNMP) and interfaces (command line interface) are in use to carry out policy provisioning in an application-dependent manner. PBM frameworks based on Java (e.g., Ponder) have used RMI (Remote Method Invocation) to carry out provisioning. However, having in mind their deficiencies [10] and the need for interoperable standards, research community and industry have been moving toward XML-based management protocols. The trend towards Web Services and XML/HTTP-based management has also affected PBM.

3 Self-management and the Autonomic Paradigm

Self-management refers to the ability of independently achieving seamless operation and maintenance by being aware of the surrounding environment [6]. It has been closely related with autonomic computing and self-maintained systems [8]. This ability is widely embedded in the natural world, allowing living organisms to effortlessly adapt to diverse habitats. Without planning or consciousness, body's mechanisms work in the background to maintain a constant temperature. To imitate nature's self-managing abilities and apply them to the management of network and systems, the latter should be provided with the logic and directives for their operation and in addition the means to sense their operating environment. Self-management has been closely related with control systems and particularly to closed-loop controllers. By using a system's output as feedback, a feedback loop allows the system to become more stable and adapt its actions to achieve desired output. From the definitions above, it is evident that two main functions are required to support self-management. These two functions are interrelated and interdependent, thus forming a closed control loop with feedback (Fig. 2):

A. Provide the logic and directives to achieve seamless operation and maintenance.
B. Provide the means to sense and evaluate their operating surrounding environment.

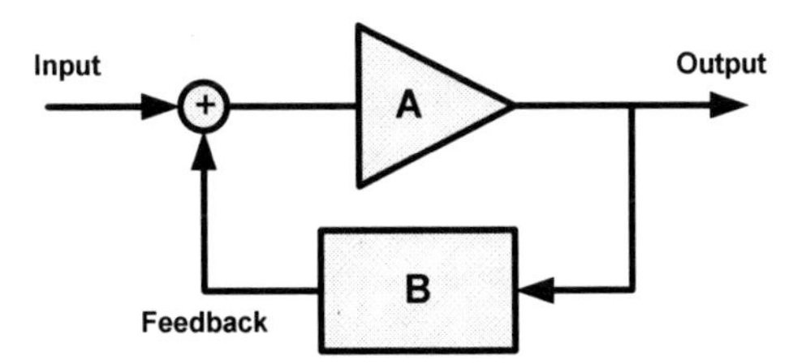

Fig. 2. Closed control loop with feedback

In 2001, an influential research declaration from IBM had introduced the concept of Autonomic Computing, which encapsulated the aspects of self-management in an architectural blueprint. The concept was inspired by the ability of the human nervous system to autonomously adapt its operation without our intervention and has appealed to researchers worldwide. IBM's vision [8] has fuelled intense research efforts both in industry and academia. In essence, autonomic computing and self-management are considered synonymous. According to IBM, "autonomic computing is a computing environment with the ability to manage itself and dynamically adapt to change in accordance with business policies and objectives." In addition, four quintessential properties of a self-management system were identified, frequently referred as self-* or self-CHOP properties:

- Self-Configuration
- Self-Healing
- Self-Optimization
- Self-Protection

Self-management concepts are increasingly used in research following the introduction of the autonomic manager (AM) component, as proposed by IBM. Major IT and Telco players are showing their research interest in autonomic networking and self-management, e.g. Motorola in [14] and Microsoft in [9] . In addition, intense interest is shown in autonomic network management from Academia [11]. The autonomic manager architectural component has become the reference model for autonomic and self-managing systems. It is a component that manages other software or hardware components using a control loop. The closed control loop is a repetitive sequence of tasks including Monitoring, Analyzing, Planning, and Executing functions. The orchestration of these functions is enabled by accessing a shared Knowledge base. The reference model is frequently referred as K-MAPE or simply MAPE, from the initials of the functions it performs. The use of a feedback loop raises concerns about a system's stability and according to control theory, a "valid operating region" of a feedback loop should be specified, indicating the range of control inputs where the feedback loop is known to work well [9]. Based on the definition of autonomic management, policies are identified as the basis of self-managing systems, encapsulating high-level business objectives.

Research on autonomic systems has been intense during the past years, aiming to embed the highly desirable self-managing properties to existing and future networks. The roadmap to autonomic management is indicative of a gradual evolution and can be used to evaluate a system's progress. Accordingly, management frameworks can advance through different maturity phases before becoming autonomic:

- Basic: manually operated management operations
- Managed: management technologies used to collect and synthesize information
- Predictive: correlation among management technologies provides the ability to recognize patterns, predict optimal configuration and suggest solutions to administrators
- Adaptive: management framework can automatically take actions based on available knowledge, subject to the supervision of administrators
- Autonomic: business policies and objectives govern infrastructure operation. Users interact with the autonomic technology tools to monitor business processes and/or alter the objectives

Apparently the road to self-management is long and a series of issues will need to be resolved on the way. Until now, a complete self-management solution is not available. Instead, researchers and practitioners have attempted to partially tackle self-management by implementing some of the desired properties and adopting a gradual transition. Each of the four desired capabilities is contributing to the overall goal of enabling truly self-managed systems.

4 Distributed Policy-Based Management for Mobile and Wireless Networks

The principles of autonomic management and self-organization are envisioned by researchers as a natural path for the Future Internet. Autonomic management is expected to simplify the painstaking tasks of managing complex large-scale systems through the use of automated closed-loop management. Therefore, rather than proposing a holistic framework for the Future Internet, the proposed approach concentrates on the design of those architectural entities that would facilitate autonomic wireless networking. Following an evolutionary path, each autonomic support entity (ASE) is designed as an extension of the ongoing Future Internet design, maintaining backward compatibility with today's Internet. Based on open standards and interfaces, this approach should ensure the interoperable and future-proof integration of ASEs with the evolving design of the FI.

Three cooperating autonomic support entities (ASE) are proposed for the enhancement of wireless and mobile networking in the FI:

1. Decentralization (DCN.ASE)
2. Mobility (MOB.ASE)
3. Ad hoc communications (AHC.ASE)

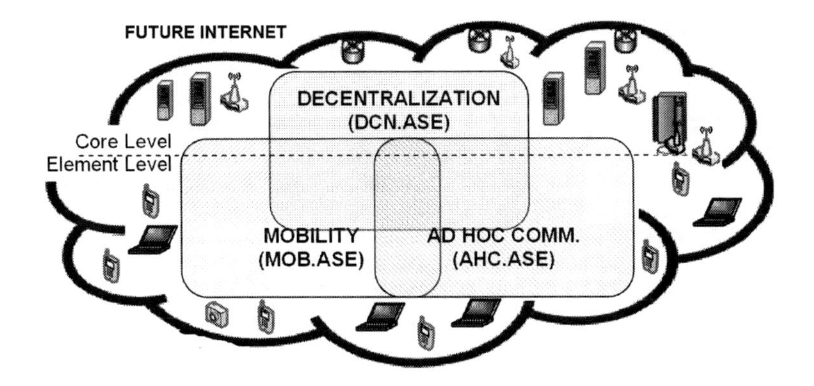

Fig. 3. Autonomic Support Entities for the Future Internet

As shown visually in Figure 3, these entities will operate and cooperate at different levels. They are designed address the lack of management decentralization (DCN.ASE), seamless mobility (MOB.ASE), and spontaneous communications (AHC.ASE) from today's Internet. The separation between Core and Element levels indicates the separation of functionality between the operator-owned network core and the user-owned devices respectively.

The motivation behind the Decentralization ASE is to enable collaborative management of large-scale networks, focusing on mobile and wireless access networks and their interconnection with fixed ones. A key goal is the efficient and scalable management of personal-area networks (PAN). In addition, the secure interactions of mobile and nomadic PANs with authorized or non-authorized wireless local-area networks (WLANs), as well as wide-area networks (WANs), will need to be addressed. The functionality of this entity is based on the distributed policy-based management (DPBM) paradigm and spans equally between core and element levels.

This means that part of the entity's functionality will be hosted in the network core, e.g. policy definition, conflict resolution and business plan implementation. However, an important part will be hosted on network elements, i.e. devices owned by end-users. Such functionality will include policy distribution, local management activities, user preference enforcement and privacy protection. For the latter end-user functionality, the Decentralization ASE will cooperate with the Mobility and Ad hoc communication ASEs, provisioning them with the appropriate policies that guide their autonomic behavior. The Mobility ASE will enable the seamless connectivity of users between different access networks and different devices as well, while the Ad Hoc Communications ASE will cater for the users' need for spontaneous communications. The cooperation of these entities is enabled though the Distributed PBM infrastructure (Figure 4), which decentralizes management tasks and pushes intelligence closer to the network edge. This can result in quicker reaction times and policy decisions for users, thus resulting in more stability. For more details on the implementation and evaluation of DPBM, the reader is referred to [7] and references within.

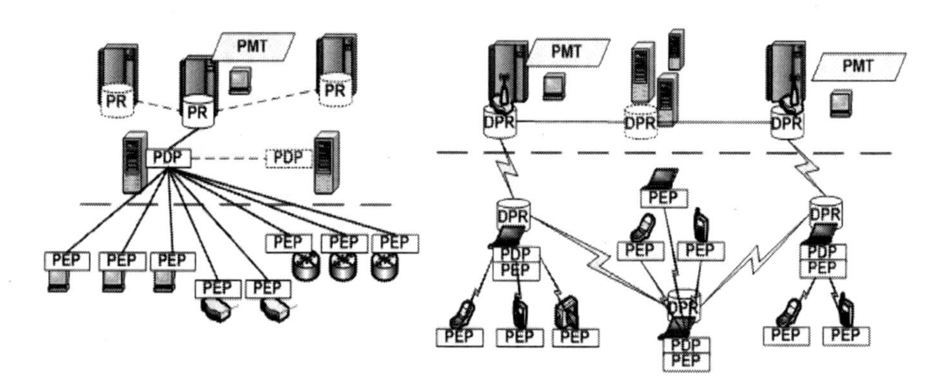

Fig. 4. Centralized versus Distributed PBM

5 Open Issues

While autonomic management is gradually implemented in today's networks, there are several open issues that need to be resolved. We focus on the two most prominent, indicating the scope for further examination and future work.

- Centralized vs. distributed control: the architecture of PBM systems is predominantly based on a centralized or hierarchical paradigm, following the organization of the managed networks. As a result, the majority of PBM functionality and protocols follow these paradigms, e.g. the manager-agent model for policy provisioning and the centralized policy repository storage. To enable distributed PBM, the coordination of multiple policy decision points (PDP) needs to be addressed in combination with decentralized policy storage and provisioning. Recently, the emergence of highly distributed computing systems (cloud computing) has motivated the decentralization of policies and their distributed management. Departing from centralized PDPs deployment, the distributed control of multiple PDPs need to be investigated.

- Conflicting policies and policy analysis: policy refinement is the process of deriving a concrete policy specification from higher-level objectives or goals [12]. It is an important process that leverages the potential of PBM frameworks, therefore it has received significant research interest, aiming to provide automated solutions [1]. The process is further hampered by the risk of producing inconsistent policy specifications, giving rise to concerns about policy conflicts and the need for policy analysis. The need for policy analysis and the lack of tested solutions is one of the main drawbacks of policy-based systems. Policy analysis [3] refers to the examination of policies and the verification of their current and future consistency. In complex environments where a number of policies need to coexist, there is always the likelihood that policies may conflict, either because of a specification error or because of application-specific constraints. It is therefore important to provide the means of detecting conflicts in the policy specification.

In spite of obstacles, the benefits from implementing autonomic and policy-based management solutions are significant. The industrial interest in standardization activities of 3GPP, like Self-Organizing Networks (SON) [18],[19] and Policy Charging and Control (PCC) [17] architecture, illustrate the potential of the described concepts and open new avenues for R&D.

Acknowledgements. This work is supported by the Cyprus Research Promotion Foundation's Framework Program for Research, Technological Development and Innovation 2009-2010 (DESMI 2009-2010), co-funded by the Republic of Cyprus and the European Regional Development Fund, and specifically under Grant ΔΙΔΑΚΤΩΡ/0609/21.

References

1. Bandara, A.K., et al.: Policy refinement for IP differentiated services Quality of Service management. IEEE Transactions on Network and Service Management 3(2), 2–13 (2006)
2. Boutaba, R., Aib, I.: Policy-based Management: A Historical Perspective. J. Netw. Syst. Manage. 15(4), 447–480 (2007)
3. Charalambides, M., et al.: Dynamic Policy Analysis and Conflict Resolution for DiffServ Quality of Service Management. In: 10th IEEE/IFIP Network Operations and Management Symposium, NOMS 2006, pp. 294–304. IEEE (2006)
4. Clemm, A.: Network Management Fundamentals. Cisco Press (2006)
5. Flegkas, P., Trimintzios, P., Pavlou, G.: A policy-based quality of service management system for IP DiffServ networks. IEEE Network 16(2), 50–56 (2002)
6. Hadjiantonis, A.M., Pavlou, G.: Policy-Based Self-Management in Wireless Networks. In: Context-Aware Computing and Self-Managing Systems, pp. 201–272. CRC Press (2009)

7. Hadjiantonis, A.M., Pavlou, G.: Policy-based self-management of wireless ad hoc networks. In: IFIP/IEEE International Symposium on Integrated Network Management, IM 2009, pp. 796–802 (2009)
8. Kephart, J.O., Chess, D.M.: The vision of autonomic computing. Computer 36(1), 41–50 (2003)
9. Mortier, R., Kiciman, E.: Autonomic network management: some pragmatic considerations. In: ACM Proc. 2006 SIGCOMM Workshop on Internet Network Management, INM 2006, Pisa, pp. 89–93 (2006)
10. Pavlou, G.: On the Evolution of Management Approaches, Frameworks and Protocols: A Historical Perspective. J. Netw. Syst. Manage. 15(4), 425–445 (2007)
11. Pavlou, G., Hadjiantonis, A.M., Malatras, A. (eds): D9.1:Frameworks and Approaches for Autonomic Management of Fixed QoS-enabled and Ad Hoc Networks. EMANICS Network of Excellence, Deliverable (2006), `http://www.emanics.org`
12. Sloman, M., Lupu, E.: Security and management policy specification. IEEE Network 16(2), 10–19 (2002)
13. Strassner, J.: Policy-Based Network Management: Solutions for the Next Generation. Morgan Kaufmann (2003)
14. Strassner, J., Raymer, D.: Implementing Next Generation Services Using Policy-Based Management and Autonomic Computing Principles. In: 10th IEEE/IFIP Network Operations and Management Symposium, NOMS 2006, pp. 1–15 (2006)
15. Verma, D.: Policy-Based Networking: Architecture and Algorithms. Sams (2000)
16. Verma, D.C.: Simplifying network administration using policy-based management. IEEE Network 16(2), 20–26 (2002)
17. 3GPP specification: 23.203 Policy and charging control architecture, `http://www.3gpp.org/ftp/Specs/html-info/23203.htm`
18. 3GPP specification: 32.500 Telecommunication management; Self-Organizing Networks (SON); Concepts and requirements, `http://www.3gpp.org/ftp/Specs/html-info/32500.htm`
19. 3GPP specification: 32.541 Telecommunication management; Self-Organizing Networks (SON); Self-healing concepts and requirements, `http://www.3gpp.org/ftp/Specs/html-info/32541.htm`

On the Provision of Advanced Telecommunication Services in Rural Areas

Lambros Lambrinos

Department of Communication and Internet Studies, Cyprus University of Technology
lambros.lambrinos@cut.ac.cy

Abstract. Even though in many parts of the world people enjoy the features of modern telecommunication services, a significant number of inhabitants in rural areas are deprived of access to such facilities. In this text we propose how we can utilize existing technologies for the provision of telecommunication facilities in rural areas: IP telephony based upon open source software and off-the-shelf hardware and voice messaging based upon the Delay Tolerant Networking paradigm.

Keywords: VoIP, delay tolerant network, voicemail, rural telecommunications.

1 Introduction

In the current era people take advanced telecommunication services for granted and assume that such facilities are available worldwide. Telecommunication operators are not prepared to invest in areas that are not profitable for them with prime examples being rural places with a small number of inhabitants or places that are remote and infrastructure development costs are high. As a result, there are many areas in the world lacking access to the latest telephony and internet technologies and when those technologies are provided, their cost may not be affordable for most of the potential users residing in such areas due to socio-economic factors.

In the context of this work, there is particular interest in dealing with two scenarios: in the first, there exist many situations where advanced telephony (e.g. voice-mail) and internet services are available in an area but their cost may be prohibitively high for the potential customers; in the second, there are areas that are totally disconnected and no telephony and internet services are provided at all or they are provided opportunistically and with limited functionality.

2 Advanced Telephony Features for Rural Areas

Evidently, rural areas do not always enjoy the advanced telecommunications facilities that are available in urban settings. However, during the last few years, the landscape in the telecommunications sector is changing dramatically; the large and traditional telecom operators are no longer monopolizing the market and face fierce competition from alternative carriers and cable operators. The major reason for this being data

A.M. Hadjiantonis and B. Stiller (Eds.): Telecommunication Economics, LNCS 7216, pp. 209–215, 2012.

convergence: the conversion of speech and video signals into data which is ´transported´ using the same networks used for traditional Internet traffic (for example, email, web browsing and file exchange data). Data convergence has resulted in radical changes in the telecommunications sector and has enriched Internet-based communication between users.

Our goal is to demonstrate how community-driven efforts can assist in reducing the cost of telecommunications, as well as, in providing access to features that would have not been available otherwise. With such an observation at hand, we propose here the design of a low-cost wireless network architecture (Figure 1) to support voice services over Internet based networks, i.e. develop a VoIP infrastructure for rural areas. To minimize costs, we chose to use freely available open-source soft-ware in the core of our system as explained in the sequel.

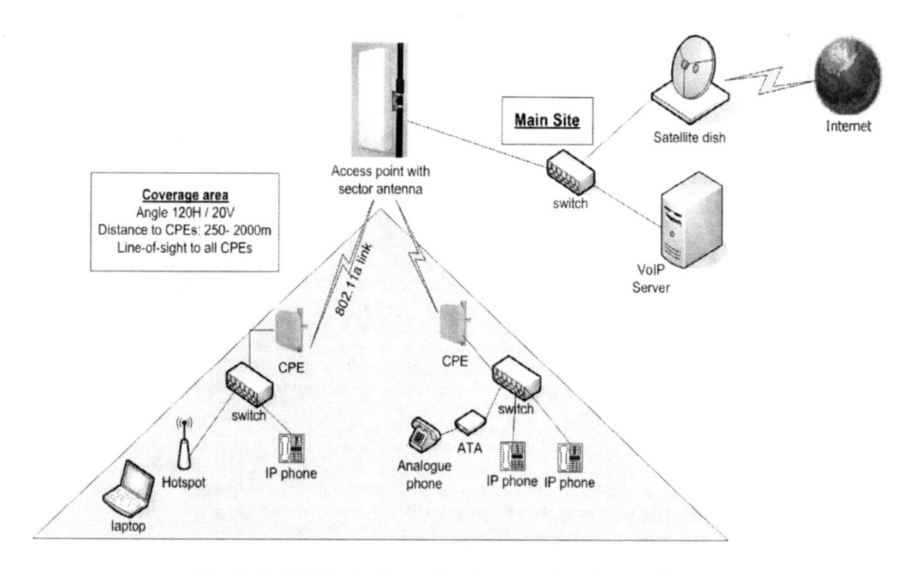

Fig. 1. A VoIP telephony implementation for rural areas

Voice and video over an IP network provide for an enriched and low-cost communication experience. Initial implementations of VoIP technology were based on proprietary protocols which gradually gave way to more open interoperability standards with the Session Initiation Protocol (SIP) [8] being widely adopted. SIP is a flexible text-based application layer signaling protocol used in the setup and termination of multimedia sessions (e.g. VoIP calls); it follows a request-response approach in its operation.

Current applications of SIP may operate based upon the peer-to-peer model (i.e. there is direct SIP-based communication between two entities) or may rely upon a SIP server which acts as a registrar holding location information and preferences for its registered users; such information is used for call establishment. As a result of its wide appeal, many end user devices (IP phones) as well as software tools (soft phones) implement SIP; along with a range of SIP compliant IP-PBXs they constitute a fully fledged telephony solution. The IP-PBX systems support a vast number of features that range from simple voice messaging to Interactive Voice Response (IVR), Conference Calling, and Automatic Call Distribution.

Our system is designed to provide enhanced telephony services in a rural community and it is based upon an open source implementation of a SIP PBX (asterisk [1]) and other low-cost technologies. Using a wireless infrastructure we interconnect our SIP-based telephony exchange with the client endpoints installed at various locations enabling seamless internal communication as well as connectivity to the normal telephone network (facilitated over analogue lines and a satellite link).

Wireless networks offer rapid, and hence low cost, connectivity solutions in scenarios where wired networks are not easily implementable; as discussed this is exactly the case in rural areas. The continuous enhancements in terms of bandwidth and link distance enable the deployment of long distance (tens of kilometers) point-to-point links between two locations in addition to the traditional hotspot deployments.

The pilot implementation of the architecture has provided some interesting findings and pinpointed some open issues. An open source IP-PBX was used and the end-points ranged from simple and advanced IP phones, adaptors for existing analogue phones to preserve user device familiarity as well as software phones (soft phones). Users welcomed the extra features provided in comparison with the plain analogue telephony they were used to; more details can be found in [9].

3 Utilizing Delay Tolerant Networks for Voice Messaging

The scenario presented in the previous section assumed that services are generally available in an area; some areas in the world do not even have a basic telephone connection and long range wireless links are used to provide rudimentary access (e.g. through a single location servicing a whole village) to the Public Switched Telephone Network (PSTN) and perhaps the Internet.

One other feature that is still quite popular in the analogue and mobile telephony world is voice messaging which is also offered by VoIP-based (i.e. packet switched) systems; one advantage is that such systems can also be configured to send the message as an email attachment to the user associated with the number called. Voicemail-to-email is primarily done in order to expedite delivery of the message (e.g. the user is not at work but has email access) but the user's email may also not be readily accessible.

Since voice messaging is by nature delay tolerant, it is well suited in scenarios where connectivity is intermittent or network resources are scarce and real-time voice communications can not be guaranteed. To realize the envisioned elastic voice messaging implementation we suggest the deployment of local infrastructures in combination with Delay Tolerant Networks (DTN) for data transfers. More particularly, we look into the provision of voicemail facilities over DTN; such a service allows users in an area where a telephony service is not regularly available (or not available at all) to communicate (through voice messages) with users connected to the PSTN.

Note, that our target application is slightly different than traditional voice messaging where message delivery takes place immediately and access time depends solely on the recipient; we have another inherent delay due to the frequent delivery network disconnectivity but on the other hand, a messaging service is provided where it would not normally be available.

Evidently to support such delay-tolerant messaging service, traditional TCP/IP protocols could not suffice as the viable approach. Instead, DTN techniques are employed that are able to combat disconnections and sustain information flow. A DTN defines a network overlay architecture [3] that delivers packets using asynchronous message forwarding. It is designed to operate in environments with limited expectations of end-to-end connectivity and node resources; to this extent, it was initially proposed to facilitate interplanetary communication but as we will see it has also been useful in other applications.

Briefly put, DTNs are based upon a store-and-forward approach for data delivery. Nodes within a DTN network act as mules; they store received messages and hold them until they "carry" them over to the next hop on their way to their destination. Before leaving its source, data is assembled into 'bundles'; the bundle protocol [2] is an application layer protocol that manages bundle propagation with the DTN area.

To interconnect a DTN with the internet infrastructure and enable the delivery and reception of bundles, a "Convergence Layer Adapter" (CLA) is used. The CLA essentially acts as an intermediary between a DTN application and the IP protocol. Different CLA implementations offer varying services to the applications or try to optimize a particular form of communication; the most common examples are the TCP Convergence Layer Protocol [4], Saratoga [10] that mimics UDP and the DTN session layer [5] which is optimized for receiver-driven applications and multicast communication.

As mentioned in the previous section, Voice-over-IP is a technology based on packet-switching that is nowadays standardized and easily accessible. Further, the Session Initiation Protocol is the established standard for setting up VoIP calls and the information it conveys during the call setup process is vital for our application. We are again using the SIP-based Asterisk PBX which is an open source solution offering advanced functionality along with access to the PSTN via special gateways; more importantly, it offers the ability to extend its call handling mechanisms through user-defined programs that can be launched on-demand.

Combining the technologies and tools just described, we built a system that utilizes voicemails to facilitate communication of users residing in isolated and disconnected geographical regions with the "outside" world. More precisely, in regions where communication through live phone calls is not feasible due to different network or capacity limitations, the proposed system records users' messages and attempts to deliver them to their intended recipients in the form of voicemails.

In order to achieve this and allow recorded messages to exit the disconnected geographical regions, the system relies on a Delay Tolerant Network infrastructure for the storage and forwarding of recorded voicemails. Thus, at the application layer, VoIP technology is used to handle the voice messages while a convergence layer is employed to link the two networking interfaces.

From the system architecture (Figure 2), we examine here a couple of major components and operations whereas more details can be found in [7].

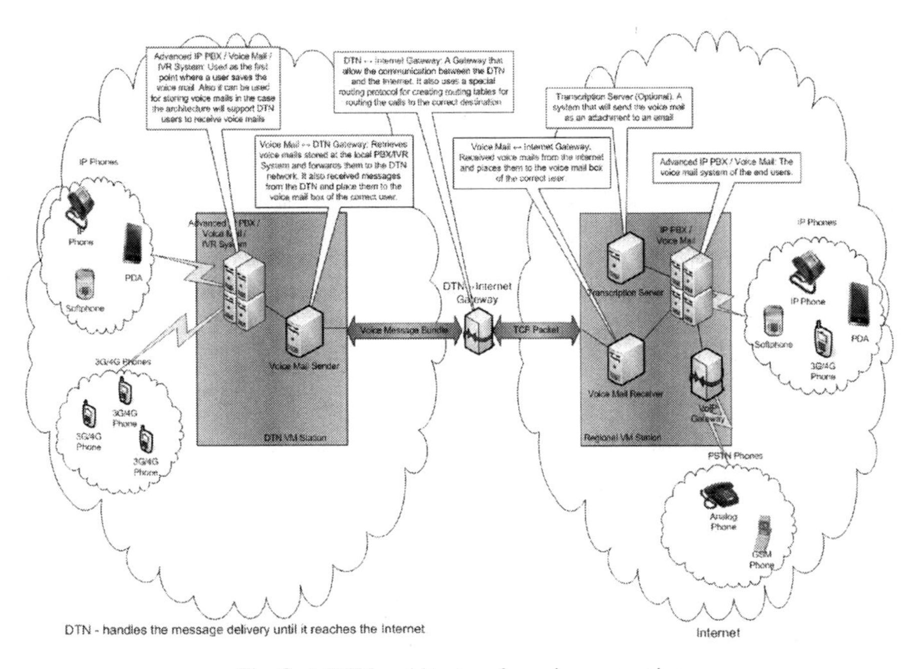

Fig. 2. A DTN architecture for voice messaging

The DTN Voicemail station is in general terms the entity responsible for processing calls originating from users inside a DTN area (i.e., disconnected region), re-cording the voice messages and initiating their delivery procedure. It consists of two modules: the IP-PBX and the Voicemail Sender.

The role of the IP-PBX is to process a user's call request; in the case that connectivity is not available (which should be the norm in a DTN region) instead of dropping the call, it should prompt the user to leave a voice message. After successfully recording the voice message, the PBX creates a special file that contains the actual message along with destination information which can be automatically extracted from the SIP INVITE message.

Once all the required information is in place, the DTN VM Station starts the process of forwarding the recorded voicemail to its intended destination over DTN. The Voicemail Sender is the component responsible for retrieving the voice messages and using their destination information to initiate their delivery. The message recipient can be a SIP user (in the form of a SIP URI) or a number in the PSTN (we refer to this information as Recipient ID).

The next step for the Voicemail Sender is to identify the destination network, i.e., the domain name or the IP of the gateway that serves the recipient. For achieving this, the Voicemail Sender utilizes the DUNDi protocol [6] and more precisely its lookup operation. If the lookup provides the destination domain, the message is sent directly towards that domain; on the other hand, if the lookup operation returns no results, the voice message is blindly forwarded to the DTN – Internet Gateway. Considering that the DTN – Internet Gateway has a permanent connection to the Internet, we assume that it has more updated DUNDi information than the Voicemail Sender (that resides in DTN region).

Extending the above functionality, the DTN Voicemail station and namely its IP-PBX component can act as a repository for incoming voice messages (i.e. for messages that arrive from other DTN regions or the Internet).

In its turn, the DTN-Internet gateway serves two purposes: the first is to provide interoperability between the Delay Tolerant Network and the Internet, and the second is to locate destination users and forward received voicemails to them. To serve the abovementioned purpose the DTN ↔ Internet Gateway processes bundles which were sent to it by a DTN VM station (namely the Voicemail sender as described above). If the Voicemail sender could not identify the destination do-main, the DUNDi protocol is used again for this purpose. Once the final destination is known, the bundle is converted to TCP/IP packets which are sent over the Internet towards their final destination. The DTN-Internet gateway also maintains a list of potential gateways for PSTN termination; one of these will be used to de-liver the message when the recipient ID is a PSTN number.

At the receiving end of a voice message, a set of components make up the Regional Voicemail station. The first point of contact is the Voicemail Receiver, the entity that receives the voice messages and starts the final delivery process by notifying the associated IP-PBX. According to the recipient-ID a number of actions are possible:

- The recipient is a user of the local IP-PBX (SIP URI); a call is automatically placed to that user's extension and once it is answered the message is played back. If there is no answer, a voicemail is created and deposited in the user's voicemail inbox.
- The recipient is a PSTN number; a call is automatically placed to that number via the gateway connected to the local IP-PBX. Once the call is answered the voice message is played back. If there is no answer the call is retried after some time.

Finally, it is important to emphasize that the proposed system is transparent to the end-users; the caller places a regular phone call leaving a voicemail message and eventually the callee simply receives a regular voicemail message.

4 Conclusions

In conclusion, through the two examples presented, we have demonstrated how technologies readily available can be utilized to enhance the telecommunication facilities offered in rural areas. Where coverage has already been extended, community efforts can drive down the cost of access to advanced telecommunications.

References

1. Asterisk - The Open Source Telephony Project, `http://www.asterisk.org/` (accessed October 11, 2011)
2. Bundle Protocol Specification, RFC 5050 (2007)
3. Delay Tolerant Networking Architecture, RFC 4838 (2007)

4. Delay Tolerant Networking TCP Convergence Layer Protocol, Internet Engineering Task Force Draft (2008)
5. Demmer, M., Fall, K.: The Design and Implementation of a Session Layer for Delay-Tolerant Networks. Computer Communications Journal 32(16), 1724–1730 (2009)
6. Distributed Universal Number Discovery (DUNDi), Internet Draft (2004)
7. Lambrinos, L.: Deploying Open-Source IP telephony in Rural Environments. In: 2nd Next Generation Mobile Applications, Services and Technologies (NGMAST), Cardiff, UK (September 2008)
8. SIP: Session Initiation Protocol, RFC 3261 (2002)
9. Tziouvas, C., Lambrinos, L., Chrysostomou, C.: A delay tolerant platform for voice message delivery. In: International Workshop on Opportunistic and De-lay/Disruption-Tolerant Networking (WODTN 2011), Brest, France, October 3-6 (2011)
10. Wood, L., Eddy, W.M., Ivancic, W., McKim, J., Jackson, C.: Saratoga: a Delay-Tolerant Networking Convergence Layer with Efficient Link Utilization. In: International Workshop on Satellite and Space Communications, pp. 168–172 (September 2007)

Electrical Power Systems Protection and Interdependencies with ICT

George M. Milis, Elias Kyriakides, and Antonis M. Hadjiantonis

KIOS Research Center for Intelligent Systems and Networks, University of Cyprus
{milis.georgios,elias,antonish}@ucy.ac.cy

Abstract. This section discusses the protection of electrical power systems (EPS) and its relation to the supporting Information and Communication Technologies (ICT) infrastructure. Several dimensions are addressed, ranging from the need of protection and available protection schemes to the identification of faults and disturbances. The challenges brought by recognizing the interdependent nature of today's and future's EPS and ICT infrastructures are also highlighted, based on the Smart Grid and the System of Systems perspectives.

Keywords: electrical power system, protection, critical infrastructures, disturbance, interdependencies.

1 Introduction

Our modern lifestyle is increasingly dependent on critical infrastructures, like the Electrical Power System, the Water Distribution Networks, the Telecommunication networks, the Transportation networks, even the Health sector, the Finance sector, and many more. The robust operation of these infrastructures is often taken for granted. But what if their operations were disrupted or even ceased? What would be the effects on our economy, our society, or on our health and safety? Around the globe, a number of communities have recognized the criticality of such infrastructures and have set their investigation and protection as a top priority, being actively engaged in research and development activities at national and international levels.

According to the European Commission [COM(2005) 576] [5] and the European Programme for Critical Infrastructure Protection (EPCIP), a "Critical Infrastructure (CI) is defined as an asset, system or part thereof located in Member States which is essential for the maintenance of vital societal functions, health, safety, security, economic or social wellbeing of people, and the disruption or destruction of which would have a significant impact in a Member State as a result of the failure to maintain those functions". The term Critical Infrastructure Systems (CIS) includes the public, private and governmental infrastructure assets, and interdependent cyber and physical networks. As already mentioned, due to their outmost importance to society, existing and future CIS need to be protected from disruption and destruction. Several threats need to be considered, including among others natural disasters, accidental damage, equipment failure, human error, and terrorist attacks. According to Council Directive [2008/114/EC] [6], CIS

A.M. Hadjiantonis and B. Stiller (Eds.): Telecommunication Economics, LNCS 7216, pp. 216–228, 2012.

protection should be based on an all-hazards approach, recognizing the threat from terrorism as a priority. The vast range of threats makes the protection of critical infrastructures a highly complex and interdisciplinary research domain. For example, the issue of interdependencies among critical infrastructures needs to be investigated to prevent propagation and cascading of failures between infrastructures. The research communities' understanding of interdependencies is still immature and requires the coordinated involvement of several different disciplines. On top of that, the increasing reliance on telecommunication networks and the Internet for communication, poses novel threats to critical infrastructures and there is an urgent need to prepare for cyber-attacks.

This section focuses on Electrical Power Systems (EPS) where the majority of economic activities depend upon, as well as the operation of many other infrastructures. In essence, an EPS is one of the most critical infrastructures with its protection being a priority worldwide. The section further discusses the interdependence between the EPS and the ICT infrastructures, presenting open problems and possible solutions.

2 The Need to Protect Electrical Power Systems

Electrical Power Systems comprise of large and complex machinery and assets that are responsible for the generation of electric power in sufficient quantity to meet the present and estimated future demands of the customers, its transmission to the areas where it will be consumed and its distribution within those areas to the actual demand sites. This process is synchronous and continuous and is a fundamental asset of our modern society and economy. Therefore, any disruption of its operation has huge negative effects of unforeseen size (unbalances and people reactions, disruption of economic activity, etc.) [11]. However, the EPS are subject to disturbances in their normal operation which subsequently cause 'faults', that occur either due to internal technical failures or due to external causes [15],[20]. Therefore, the protection of the EPS is defined as the effort to ensure its reliable operation within an environment with effects of disturbances, failures and events that put the system at risk. More specifically, it is necessary to ensure: i) public safety, ii) equipment protection, and iii) quality of service, by limiting the extent and duration of service interruption, as well as, minimizing the damage to the involved system components. A non-exhaustive list of EPS components that need to be protected, would include the following [10],[18]:

- Human personnel.
- EPS equipment (transmission/distribution lines, generators, transformers, motors, bus-bars).
- Customer owned equipment.
- Operational integrity of EPS (ability to deliver power, stability, power quality).
- Customer operations (e.g., large, medium and small industry, financial services, transportation services, telecommunication and other infrastructures' services).

In any case, safety of human personnel and prevention of human injury should always take priority over service continuity, equipment damage, or economic losses. It is therefore necessary to prevent or, at least, to detect and clear any faults quickly, since the longer the EPS operates under a fault, the higher that the system will become unstable and lead to cascading outages.

The cost required for having the EPS perfectly safe or perfectly reliable (by adopting components that do not fail and that require minimum maintenance) would be prohibitive. As stated in [20], risk assessment is performed to define a trade-off for the acceptable levels of risk from disruption in association with relevant costs. That is, the cost of the protection of a system determines the degree of protection that can be incorporated into it. Figure 1 provides an overview of the cost of a major blackout that occurred in the US in 2003. The estimated economic cost of $6.8-10.3 billion over three days of disruption is alarming and would justify higher investments on the protection efforts of the EPS.

Approximate Start Time	Approximate End Time	Lost Megawatt	Duration		Cost of Blackout ($ Billion)	
		MW	Hour	MWh	Lower Bound	Upper Bound
8/14 - 4 PM	8/14 - 8 PM	61,800	4	247,200	$1.8	$2.8
8/14 - 8 PM	8/15 - 6 AM	30,900	10	309,000	$2.3	$3.4
8/15 - 6 AM	8/15 - 10 AM	15,450	4	61,800	$0.5	$0.7
8/15 - 10 AM	8/16 - 12 AM	13,200	14	184,800	$1.4	$2.1
8/16 - 12 AM	8/16 - 10 AM	6,600	10	66,000	$0.5	$0.7
8/16 - 10 AM	8/17 - 6 AM	2,000	20	40,000	$0.3	$0.4
8/17 - 6 AM	8/17 - 4 PM	1,000	10	10,000	$0.1	$0.1
Total Economic Cost					$6.8	$10.3

Fig. 1. The economic cost of the Northeastern Blackout, Aug. 2003, USA [1]

3 Types of System Disturbances and Their Effects

No matter what the criticality of the investment is, it is not technically and economically feasible to design and build an EPS such as to eliminate the possibility of disturbances in service. It is therefore a fact that, infrequently, and at random locations, different types of disturbances will occur, leading to faults incurred on the system. The term 'disturbance' means any event, unexpected or foreseen, which requires corrective action to be taken. Sudden disturbances on EPS may occur from factors external to the system itself, such as weather or environment, or internal factors such as insulation failure on some component of the system. The following external factors of system disturbances/failures can be identified [12]:

- Weather: One of the main causes of equipment failures, according to a worldwide survey over several years in the 1980s: about 20% of failures were attributable to weather conditions.
- Sudden changes in balance between demand and generation: They can result from numerous causes: loss of transfers from/to external systems, transmission circuit

tripping, isolating parts of the system with embedded generation or demand, etc. This type of disturbance is rather frequent, as many other types of disturbances may result in some imbalance during their development.

- Human error: errors of the personnel of a utility may occur at all stages, from planning/design to plant manufacture to plant installation to maintenance/testing to operation. In addition, members of the public may be involved in system errors unintentionally (e.g., kite flying) or consciously (e.g., illegal entry into or attack to parts of the system).

The disturbances caused by the above factors can be categorized as either faults (e.g., breakdown or burning of insulation) or abnormal operation (at local and/or system level, e.g., overload or frequency variation). Each disturbance requires protective actions and measures to minimize its impact, especially the triggering of a cascaded sequence of events.

4 Protection Measures and Devices to Minimize the Impact of Disturbances

A number of measures (including automatic mechanisms and "defense plans") should be taken in the management, planning and operation phases of EPS to minimize the effects of disturbances [13]. Protection measures often contain three main elements:

- Detect a possible disturbance and its type;
- Assess the best way to prevent it or minimize its effect at the extent possible; and
- Restore normal operation.

In general, measures are taken with the help of hardware and/or software components. However, human involvement in decision making and implementation of protection processes is equally important. The human involvement in decision making is even more important during the operational phase, particularly during the development of a disturbance. This creates a critical need for a reliable and fast communication infrastructure, which will timely convey the collected data from the equipment and distribute corrective commands to networked controllers.

4.1 Measures in the Planning, Operational, and Restoration Timescale

The system engineers are the ones defining the protection plans, as well as, what specific protective and control measures will be needed during system operation. Some of the measures (for instance, under-frequency load shedding), are addressed in the planning phase, while others (e.g. inter-tripping schemes with local impact, the necessity of which emerges at short notice) are addressed directly in the operational plan. In the event of a system disturbance, the following actions are taken, based on the operational memoranda and procedures and any special protection schemes or coordinated defense plans employed by the plant engineers: i) open breaker and under frequency relays (isolated fault), ii) switch capacitors, iii) load shedding/disconnection, iv) islanding, v) no action, vi) alarm signal, vii) generation margins adjustment, viii) demand adjustment.

As EPSs become larger and more complex, timescales for response to disturbances are shrinking, calling for the implementation of more automated measures. Effective automation requires the collection and analysis of current and voltage measurements at remote terminals of a line in order to allow remote control of equipment, thus requiring enhanced ICT infrastructure to support real-time operations.

4.2 Measures in the ICT Facilities

The EPS need very secure (physical and logical) communication links for data and speech. Such links are either system-wide or local, between substations. A way to improve the security of communications is by avoiding as much as possible the dependence on any single external provider. Usually, a utility possesses its own communication channels for connecting its equipment (e.g., power line carrier or fiber) or channels are leased from external providers, such as public communications networks or even other industries with widespread communication networks, such as railways. It is also a common practice in EPS to duplicate (at least) the SCADA (Supervisory Control and Data Acquisition) systems and the EMS (Energy Management System) within one Control Center building. In some cases, the utilities go further and provide backup of whole Control Centers in different geographic locations.

4.3 Protection System Devices and Components

In order to safeguard the robust operation of the EPS and avoid costly partial or even full blackouts, a number of sophisticated protection devices is installed and instrumented. The most important ones are mentioned below [8],[10]:

- Fuses: the most common and widely used protective device in electrical circuits. Fuses are independent, self-destructing components, aiming to isolate faulty parts of the system and save the equipment on the healthy part.
- Instrument transformers: They continuously measure the voltage and current at several locations/components of the EPS and are responsible to give feedback signals to the relays to enable them to monitor the status of the system and detect abnormal conditions in the operation.
- Circuit breakers: They are basically switches to interrupt the flow of current and they open on relay command. They are used in combination with relays, to detect and isolate faults, where fuses are unsuitable. Their important characteristics from a protection point of view are the speed of reaction and the capacity of the circuit that the main contacts are capable of interrupting.
- Tripping batteries: they give uninterrupted power source to the relays and breakers that is independent of the main power source being protected. They have an important role in protection circuits, as without them relays and breakers will not operate, making the performance of the whole network unacceptable.
- Relays: they convert the signals from the monitoring devices/instrument transformers (mostly voltages and currents) and give instructions to open a circuit

under faulty conditions or to give alarms when the equipment being protected is approaching its capacity limits. This action ensures that the remaining system remains untouched and protects it from further damage.

With the advancement in digital technology, the use of microprocessors and the availability of reliable communication infrastructures, the relays became able to monitor various parameters, which give complete history of a system during both pre-fault and post-fault conditions. This led to the name of Intelligent Electronic Devices (IEDs), which is considered the component of the future protection of EPS.

4.4 Intelligent Electronic Devices (IED)

The IEDs [8] comprise the second generation of microprocessor relays that are, in addition, utilized as data acquisition units and for the remote control of the primary switchgear. This utilization of the relays has been inspired by the fact that faults do not happen so often and in all parts of a system, therefore, relays can serve other duties as well, apart from their main protection function. In addition, utilizing the protection relays also for data acquisition and control, it achieves integration and better interoperability of the various systems such as protection, supervisory control and data acquisition.

Furthermore, the use of optical fibers in digital communications allowed the exchange of information between the relay and the substation control level to be much more reliable. The following information is typically available from the relay: i) measurement data of current and voltage, ii) information stored by the relay after a fault situation, iii) relay setting values, iv) status information on the circuit breakers and isolators, v) event information. The communication link to the relay can also be used for control purposes, i.e. i) circuit breaker open/close commands, ii) remote reset of the relay or auto-reclose module, iii) changes to the protective relay settings.

The functions of a typical IED can be classified into protection, control, monitoring, metering and communications. Specifically, the communication capability of an IED is one of the most important aspects of modern EPS and protection systems. The communication can be power line carrier, microwave, leased phone lines, dedicated fiber, and even combination of these. There is a strong need to have adequate back-up in case communication is lost, e.g. as a result of a disturbance.

An efficient combination of IEDs, programmable logic controllers (PLCs) and Remote Terminal Units (RTUs) to monitor and communicate could lead to a good level of EPS substation automation in order to improve customer service. An RTU is a useful addition to record all the information about the various parts of the EPS. Adding RTUs at the substation level, further allows the built-in intelligence to be moved to the substation level. That intelligence reduces the amount of data that must be communicated between substations and the main station. For example, information can be retrieved as and when needed from databases maintained at the substation.

5 The Electrical Power System as an Interdependent Infrastructure

Today's EPS consist of several and heterogeneous components, all connected through complex electrical networks/grids [7]. The trend in recent years is that private and public EPS (utilities) operate in interconnected power grids, thus increasing the reliability of the whole system but also generating market opportunities. This interconnection evidently improves the reliability of each member utility because any loss of generation can be transparently covered by the neighbor utilities. On the other hand, this interconnection increases the complexity of the EPS. Then, the concepts of protection, security and reliability of service become much more significant, for both individual utilities and the interconnected EPS.

Moreover, the technological advances and the necessity for improved efficiency in several societal operations resulted in increasingly automated and interlinked heterogeneous infrastructures, with consequences on increased vulnerabilities to accidental and human-made disturbances. What may appear as different parts of our societies, does indeed depend on and influence each other. E.g. an apparently irrelevant event like a small social conflict, a thrown cigarette or a delayed disposal of waste can, under similar conditions, either vanish without any significant impact or trigger riots, forest fires or epidemics. Therefore, when studying complex systems, one must consider also their interdependencies with other systems.

There are four different types of interdependencies of critical infrastructure systems, as identified in [17]:

- **Physical** interdependency: arises from a physical linkage between the inputs and outputs of two infrastructures. For example, an output of one infrastructure is required as input to another infrastructure for it to operate;
- **Cyber** interdependency: the state of an infrastructure depends on information transmitted through the ICT infrastructure (computerization and automation of modern infrastructures and widespread use of supervisory control and data acquisition (SCADA) systems);
- **Geographical** interdependency: a local environmental event can create state changes in all involved infrastructures; implies close spatial proximity of the components of different infrastructures; and
- **Logical** interdependency: the state of each infrastructure depends on the state of the other via a mechanism that is not a physical, cyber, or geographic connection. For example, various policy, legal, or regulatory regimes can give rise to logical linkage among two or more infrastructures.

Modeling the interdependencies among interlinked infrastructures and assessing their impacts on the ability of each system to provide resilient and secure services are of high importance. Specifically for the case of the EPS, following such interdependency analysis is of utmost importance so as to take steps to mitigate any identified vulnerabilities and protect the system's operation from any internal or external threat.

The first step in analyzing interdependencies is identifying them. The case presented below offers an illustration of the type of obvious and not so obvious interdependencies between the EPS and the ICT infrastructure.

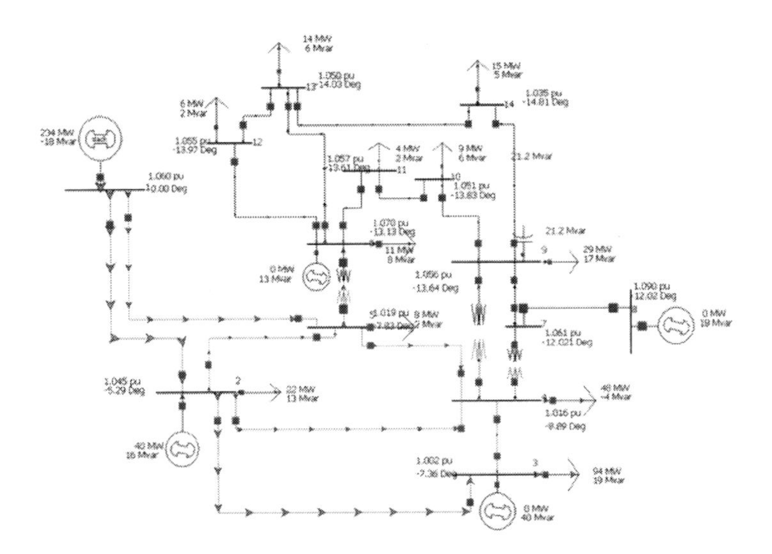

Fig. 2. The IEEE 14 bus test system

The IEEE 14 Bus Test System is a reference EPS which is extensively used in the literature for examining new concepts in a comparable way. Though not a realistic EPS, it is a miniature of real systems useful for educational purposes. It includes 14 buses, i.e. points of electric power exchange, 5 generators and several loads, i.e., the consumers of power. For illustration purposes, the 14 bus system is reorganized and characterized in seven geographic areas, and overlaid with eleven Remote Terminal Units (RTUs) to control equipment on all buses. RTUs are the computerized front-ends of IEDs and are connected using communications infrastructure with the primary control center for Supervisory Control and Data Acquisition (SCADA). The SCADA system is the communications backbone of every modern EPS.

As seen in Figure 3, a wide geographic area is considered where the EPS operates and in Figure 4 a possible network topology of communication links between RTUs in those areas is illustrated. This imaginary case assumes the EPS operator owns the communication links at the bottom half of the figure (operator network, red lines), e.g., a private optical fiber network. The six RTUs, as well as the primary and secondary control centers are directly controlled by the EPS operator. For resilience, the EPS operator leases a number of communication links from a carrier (shown with dashed red lines). The remaining five RTUs are assumed to be located in residential and industrial areas, out of the reach of the private optical fiber network of the EPS operator. Therefore, their RTUs are connected with the SCADA system using an ISP's access network.

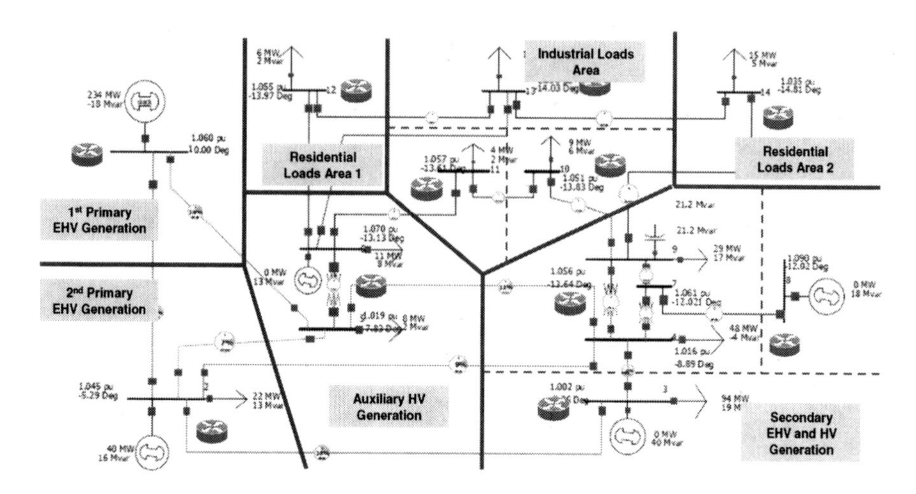

Fig. 3. Illustration of geographic areas based on IEEE 14 bus test system

Physical and **Geographical** interdependencies are relatively straightforward to identify, but require a thorough knowledge of all infrastructures involved. Since information exchange between infrastructure operators is limited, several Physical and Geographical interdependencies go unnoticed until a disruptive event occurs. A recent event in Ireland highlights such hidden interdependencies. In this case, a single transformer fault at the local EPS utility brought down the "cloud infrastructure" of two major ICT infrastructure providers all over Europe[1]. In the above simple case, the reliance of the EPS operator to external communication networks obviously creates a **cyber interdependency**, which needs to be carefully engineered to avoid future contingencies. Although the provisioning of leased communication lines from a carrier network may be tied to strict contractual agreements for availability, these agreements remain private contracts. EPS operators have expressed concerns that these agreements may not be sufficient to guarantee public safety, especially in the event of major catastrophes and natural disasters. The type of **logical** interdependencies are harder to identify and even harder to protect from. They may involve business interests and regulatory conformance of one infrastructure, which may indirectly affect the operation of another. For example, company policies may prevent a fixed-access ISP to announce performance degradation events, such as temporary network congestion. Although the EPS operator could benefit from this knowledge and proactively enable alternative communication means for isolated RTUs (e.g., via GPRS/3G data links), the lack of information exposes the EPS operator to the possibility of communication degradation and sudden disruption.

In the presented test case, the simplicity of the EPS and ICT network immediately reveals their interdependencies. However, in real EPS grids with hundreds of busses, the complexity of the interconnected ICT networks and their dependencies are

---

[1] See `https://www.datacenterknowledge.com/archives/2011/08/07/lightning-in-dublin-knocks-amazon-microsoft-data-centers-offline/`

daunting. As discussed in the next subsection, the advent of the Smart Grid is amplifying communication requirements for modern EPS, further increasing the importance to investigate them in parallel to a resilient ICT infrastructure.

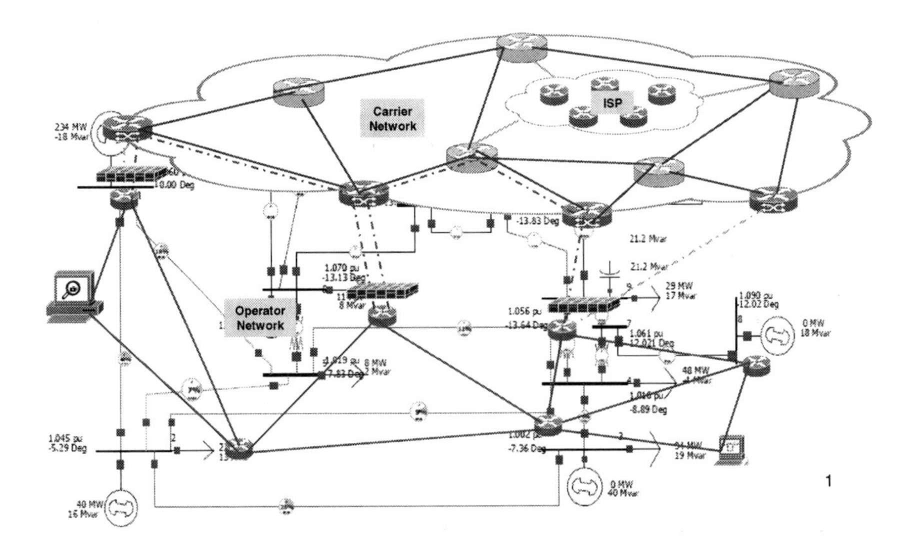

Fig. 4. SCADA system's interdependent ICT infrastructure

6 The Smart Grid and the Systems of Systems Perspective

The previous subsection presented the interdependent nature of the EPS and the need to model all types of interdependencies in order to ensure effective monitoring, control and thus protection of the infrastructure and its objectives. This fact became much more evident with the emergence of the Smart Grid concept and technologies [1], which brings the interdependencies between the EPS and the ICT infrastructures in the forefront. It has been already made clear that to ensure the effective protection of the EPS, the following three elements are considered fundamental: measurements (data defining the state of the system), data processing, and control [8]. Evidently, these elements are tightly interconnected with the ICT infrastructure, which, as seen, is increasingly employed to provide the necessary two-way communication between EPS elements, as envisioned for the Smart Grid. According to IEEE, the Smart Grid is seen as a large and complex system where different domains are expanded into three foundational layers: (i) the Power and Energy Layer, (ii) the Communication Layer and (iii) the IT/Computer Layer. Layers (ii) and (iii) are enabling infrastructure platforms of the Power and Energy Layer that makes the grid "smarter" [9].The viewpoint of IEEE is representative of the awareness around the interdependent nature of the EPS and the ICT infrastructure.

The Smart Grid calls for computers, communication, sensing, and control technologies to operate in parallel towards enhancing the reliability, minimizing the costs and facilitating the inter-connection of new sources in the EPS. The mobilization of the above ICT-enabled technologies needs to happen across broad temporal, geographical and industry scales, so as to close loops where they have never been closed before. The following is a non-exhaustive list of advantages promised by the Smart Grid technologies while being employed in the EPS:

- Self-healing from power disturbance events
- Enabling active participation by consumers in the response of the system (effective demand management)
- Operating resiliently against physical and cyber attacks
- Accommodating all generation and storage options (integration of intermittent renewable energy sources)
- Optimizing of assets and efficient operation (secure and real-time two-way power and information flows)

However, the optimization of EPS operations and resources comes with the cost of increased risk from cyber-attacks, which signifies the need to employ full cyber-security. For example, smart meters (the new metering devices that are deployed in customers' premises) are an attractive part for attacks. Malicious access to the communication channels could cause catastrophic effects on the EPS by driving decisions based on inaccurate data [16],[19].

Beyond the security threats, other aspects that need to be carefully addressed by the introduction of two-way communication are some key privacy concerns. That is, the "smart infrastructure" will require storing personal profiles of customers, to enable the determination of consumer energy behavioral patterns. Moreover, the energy consumption will be recorded remotely and in real-time, thus generating information on whether people use specific facilities.

6.1 System of Systems Architectures

A way forward to achieve real control over the operation of the EPS, thus meeting the needs of the uncertain future, seems to be the driving of the evolution in Systems of Systems (SoS) architectures [3]. An SoS is defined as a collaborative set of systems in which component-systems i) fulfill valid purposes in their own right and continue to operate to fulfill those purposes if disconnected from the overall system, and ii) are managed in part for their own purposes rather than the purposes of the whole [14].

As mentioned earlier, the Smart Grid evolves to include devices that were not previously considered, such as distributed energy sources, storage, electric vehicles and appliances. Such devices comprise heterogeneous systems that serve as integrated components of the emerging EPS and that have different characteristics, requirements for security, requirements or fault detection, protection and metering. Therefore, the traditional architectures turn to be inefficient due to the increasing demand for greater control of energy usage. The challenge is to build the grounds for an evolving SoS architecture, which will be based on open standard services mechanisms. Such architecture will avoid any hard assumptions made at the design phase, allowing a loose coupling for:

- Components to be added, replaced or modified individually in case of malfunctions, without affecting the remainder of the system;
- Components to be distributable; and
- Defining interfaces using standard metadata for application developers for use in replacing components.

The evolution of the EPS as a SoS, with the emergence of clear interfaces among component-systems and processes is expected to enhance the ability of controlling and protecting the EPS. Knowing the components and their interactions will definitely increase our ability to detect problems (created by accidental or malicious intervention), characterize them and address them quickly and efficiently.

7 Conclusions

Evidently, the operation of the EPS affects our everyday life, across many dimensions, from economical to societal. Therefore, electric utilities allocate a justifiable part of their budget for installing, operating and maintaining a sophisticated protection layer. As shown, the protection system undergoes significant evolutionary changes through the years, namely the technological advances utilized in protection devices, the algorithms that allow for the orchestration of the protection devices become more intelligent, and the protection processes become more and more automated, utilizing the ICT infrastructure. At the same time, the complexity of the resulting system drastically increases due to the amplified impact of the EPS to other critical infrastructures in our society, as well as, due to the increase of the impact of other infrastructures to the operation of the EPS. The challenge for the academia and industry is to join forces and make sure they find the necessary balance, to keep the operation of the system reliable and with high quality of service.

References

1. Amin, M., Wollenberg, B.F.: Toward a smart grid: power delivery for the 21st century. IEEE Power and Energy Magazine 3(5), 34–41 (2005)
2. Bansari, S.: The Economic Cost of the Blackout-An issue paper on the Northeastern Blackout. ICF Consulting, Fairfax, VA, August 14 (2003)
3. Chandy, K.M., Gooding, J., McDonald, J.: Smart Grid System-of- Systems Architectures: Systems Evolution to Guide Strategic Investments in Modernizing an Electric Grid
4. Chiaradonna, S., Di Giandomenico, F., Lollini, P.: Interdependency Analysis in Electric Power Systems. In: Setola, R., Geretshuber, S. (eds.) CRITIS 2008. LNCS, vol. 5508, pp. 60–71. Springer, Heidelberg (2009)
5. Commission of the European Communities, Green Paper on a European Programme for Critical Infrastructure Protection, COM (2005) 576 Final (November 2005)
6. Council Directive 2008/114/EC, On the identification and designation of European critical infrastructures and the assessment of the need to improve their protection. Official Journal of the European Union (December 2008)
7. CRUTIAL: European Project CRUTIAL - critical utility infrastructural resilience (contract n. 027513)
8. Hewitson, L.G., Brown, M., Balakrishnan, R.: Practical Electrical Power Systems Protection. Elsevier, Burlington (2004)

9. IEEE Smart Grid Conceptual Model, accessed online, `http://smartgrid.ieee.org/ieee-smart-grid`
10. Johnson, B.K.: Electrical power system Protection and Relaying. In: Lectures of ECE525-Electrical Power System Protection and Relaying at the Department of Electrical and Computer Engineering, University of Idaho, Moscow (2010)
11. Knight, U.G.: Disturbances in Electrical power systems and Their Effects. In: Electrical Power Systems in Emergencies - From Contingency Planning to Crisis Management. John Wiley & Sons (2001)
12. Knight, U.G.: The Natural Environment-Some Disturbances Reviewed. In: Electrical Power Systems in Emergencies - From Contingency Planning to Crisis Management. John Wiley & Sons (2001)
13. Knight, U.G.: Measures to Minimize the Impact of Disturbances. In: Electrical Power Systems in Emergencies - From Contingency Planning to Crisis Management. John Wiley & Sons (2001)
14. Maier, M.W., Rechtin, E.: The Art of Systems Architecting, 2nd edn. CRC Press, London (2000)
15. McLean, I.: Certificate in Electrical power system Protection. IDC Technologies, USA, Rep. on training courses (2010)
16. Mohajerin, E.P., Vrakopoulou, M., Margellos, K., Lygeros, J., Andersson, G.: A Robust Policy for Automatic Generation Control Cyber Attack in Two Area Power Network. In: IEEE Conference on Decision and Control, Atlanta, Georgia, USA (December 2010)
17. Rinaldi, S., Peerenboom, J., Kelly, T.: Identifying, Understanding, and Analyzing Critical Infrastructure Interdependencies. IEEE Control Systems Magazine, 11–25 (December 2001)
18. Soman, S.A.: Electrical power system Protection. In: Lectures of Electrical Power System Protection at the Department of Electrical Engineering, IIT Bombay
19. Teixeira, A., Amin, S., Sandberg, H., Johansson, K.H., Sastry, S.S.: Cyber-security analysis of state estimators in electric power systems. In: IEEE Conference on Decision and Control (March 2010)
20. Wall, R.W.: Introduction to Practical Electrical power system Protection. In: Lectures of EE526: Protection of Electrical Power Systems II, Department of Electrical Engineering, University of Idaho, Moscow (2005)
21. Zima, M.: Special Protection Schemes in Electric Electrical power systems, Swiss Federal Institute of Technology, Literature survey (2002)

Towards Quantitative Risk Management for Next Generation Networks

Iztok Starc and Denis Trček

Faculty of Computer and Information Science, University of Ljubljana, Slovenia
{iztok.starc,denis.trcek}@fri.uni-lj.si

Abstract. While user dependence on ICT is rising and the information security situation is worsening at an alarming rate, IT industry is not able to answer accurately and in time questions like "How secure is our information system?" Consequently, information security risk management is reactive and is lagging behind incidents. To overcome this problem, risk management paradigm has to change from reactive to active and from qualitative to quantitative. In this section, we present a computerized risk management approach that enables active risk management and is aligned with the leading initiative to make security measurable and manageable. Furthermore, we point out qualitative methods deficiencies and argue about the importance of use of quantitative over qualitative methods in order to improve accuracy of information security feedback information. Finally, we present two quantitative metrics, used together in the model, and enabling a quantitative risk assessment and support risk treatment decision making.

Keywords: computer security, economics of security, risk management, security metrics, security measurement.

1 Introduction

Information security risk management is still in its early stages with regards to measuring and quantitative assessment. Currently, risk assessment is normally based on qualitative measurement and metrics. The consequent undesirable side-effect is that risk assessment cannot provide answer to questions like "How safe is my information system (IS)?" and "How much safer is my IS then my competitors' IS? Decision making under such uncertainty is not effective. Currently, decision makers react on incidents rather than be proactive. This lagging reaction results in notable losses. Risk management paradigm has to change from reactive to proactive, where risks are identified, assessed and treated in time, before incident takes place. Furthermore, risk management is also about financial investment into security safeguards. Therefore, their spending should be justified as much as possible. This is possible only when decision makers have adequate information to evaluate discrepancy between desired and actual risk. Based on this information, appropriate and economically sound safeguards are implemented to reduce risk to a level acceptable for organization and stakeholders.

A.M. Hadjiantonis and B. Stiller (Eds.): Telecommunication Economics, LNCS 7216, pp. 229–239, 2012.

New research steps are presented in this subsection and are aiming towards computerized quantitative risk management for decision making support. First, we will present basic definitions and open problems in this area. Next, we will focus on risk assessment methodology and will address measurement and metrics issues. Finally, we will present technological architecture that enables reactive and proactive risk management in modern IS.

2 Basic Definitions

The normative reference, which is most relevant for risk management in IS, is ISO/IEC 27000-series standards for information security management systems. According to ISO/IEC 27000 [9] information security means preservation of confidentiality, integrity and availability of information.

- "Confidentiality is a property of system that information is not made available or disclosed to unauthorized individuals, entities or socio- and/or technical processes".
- "Integrity is a property of protecting accuracy and completeness of assets. There are many types of assets, tangible assets like (i) information, (ii) software, (iii) hardware, (iv) services, (v) people and also intangible assets like (vi) reputation".
- "Availability is a property of being accessible and usable upon demand by an authorized entity".
- Furthermore, and according to ISO/IEC 27000, information security may include preservation of other properties, such as authenticity, accountability, non-repudiation and reliability.

Concepts defined above are only meaningful in practice when they are linked to organization's assets and selected as operational requirements. Assets are valuable to organization and other stakeholders as well as to various threat agents. Therefore, an appropriate security assurance method has to be chosen in order to achieve organization's and stakeholders' confidence that assets satisfy the stated information security requirements and consequently its security policy and/or applicable law like [6]. For example, when organization provides service to its customers, a process security assurance method is chosen, like ISO/IEC 27001 [10]. Next, the method is applied to ensure that assets (including IS and IS services) conform to security requirements. In this way, correct, efficient and economically sound safeguards are implemented that protect assets from threat agents in such way that risk is reduced to a level acceptable for both organization and stakeholders. Risks have to be constantly monitored and when any risk factor changes then the process has to be repeated again in timely manner. This continuous activity is called information security risk management (risk management for short). Before we advance with risk management and its activities some additional basic terms have to be defined.

According to ISO/IEC 27000 information security risk (risk for short) means potential that a threat will exploit a vulnerability of an asset or group of assets and thereby cause harm to an organization and consequently cause harm to stakeholders. Vulnerability is a weakness of an asset or safeguard that can be exploited by a threat. Threat is a potential cause of an information security incident (incident for short).

We can now focus on risk management, which is, according to ISO/IEC 27005 [12], comprised of coordinated activities that aim to direct and control an organization with regard to risk. First, organization's business context has to be established that is foundation for further activities. Within this context, risks have to be identified. Next, risk assessment takes place where risk are qualitatively or better quantitatively described and prioritized against organization's risk evaluation criteria. Subsequently, risks have to be treated to achieve security requirements and correct, efficient and economically sound means of managing risk have to be implemented[1]. These means are called safeguards[2] which include policies, processes, procedures, organizational structures, and software and hardware functions. Depending upon safeguards objectives risk treatment can be accomplished in four different ways: (i) risk reduction, (ii) risk retention, (iii) risk avoidance and/or (iv) risk transfer. Finally, if risk treatment is satisfactory then any residual risk is accepted. Risk management is continuous "Plan-Do-Check-Act" process, because risk factors may change abruptly and this may lead to undesirable consequences. Thus, risk and safeguards need to be monitored, reviewed and improved, when necessary.

3 Open Problems

Information security researches are facing challenge, because current risk management practice is reactive and it is lagging behind incidents. This practice has adverse impact to the level of business objectives achieved and results in huge damages due to following reasons.

- **Plan and Do Problems.** Safeguards may be not correct and/or effective enough to protect assets from harm. Software (including security software) is buggy and single attack can disable safeguards and expose assets. In addition, threat landscape is constantly changing and future threats are not anticipated in time, because (i) business contexts of organizations are changing, (ii) user dependence on ICT is rising and (iii) ICT grows in size and complexity and (iv) ICT interdependencies is increasing.
- **Check and Act Problems.** Incapability to provide answers to security and risk related questions in time means that security cannot be managed efficiently, e.g., "How secure is the organization?" or "What is the degree of information security risk?". Logical consequence of this incapability is wider window of vulnerability [16] and increased duration of asset exposure. Thus, probability of information security incident is greater on average. Eventually, answer to two questions above is provided when risk manifests itself as incident and assets are damaged. Finally, risk management reacts on this lagging (human perceptible) indication. At this (too late) point, organizations as well as stakeholders perceive that security requirement are not fulfilled and risk level is unacceptable.

---

[1] Other product assurance methods such as Systems Security Engineering – Capability maturity Model [8] and/or process assurance methods such as Common Criteria [7] are used to ensure safeguard correctness and efficiency. ISO/IEC TR 15443-2 [13] lists a comprehensive list of assurance frameworks.

[2] Safeguards are also known as controls or countermeasures. Standard ISO/IEC 27002 [11] provides a comprehensive list of safeguards.

How are these problems addressed in information security research? On one hand, new security mechanisms [1] are studied to overcome brittleness of software and to safeguard IS more effectively. In parallel with this effort, new product security assurance methods are developed to evaluated security mechanism's strength as well as process assurance method to evaluate correctness of security mechanism's design, implementation, integration with the IS and deployment.

On the other hand, no security mechanism or security assurance method seems to be perfect to this point. Risk factors may change abruptly and ISs are changing, so statistically relevant long-term data is not available to enable security forecasting and information security insurance practical. Therefore, risk has to be reduced and this means safeguards have to be constantly monitored, reviewed and last but not least, improved, when necessary. This is possible only if information security feedback is accurate and is provided in real-time. Only then, decision makers have adequate information.

Detection of security precursor before incident takes place and risk forecasting ability is a research priority. In order to accomplish this, better security measurements methods have to be defined that are (i) accurate, (ii) real-time, (iii) economically sound, and (iv) measure security attributes according to business requirements. Security attribute measurement takes place on various IS objects, e.g., on routers, workstations, personal computer, etc. Acquired raw data can be then interpreted using metrics/indicators that are in fact analytical models, which take basic measurements as an input and return organization's information security state. This feedback information should be provided to decision maker as soon as possible in order to enable pro-active risk management rather than reactive. Thus, leading indicators should be chosen over lagging indicators, to prevent incident rather than to detect and manage incidents.

The indicator output is manually or computationally compared to organization's own risk evaluation criteria and risk management action is taken if necessary. Using described measurement and metrics as a foundation, security research aims to create also self-adapting security information and event management systems (SIEM) [2] that take actions based upon indicators values and measurements.

Before we advance towards computerized risk assessment for proactive risk management, we will analyze current risk assessment practices as well as address security metrics and measurement issues and provide some problem solutions.

4 Current Risk Assessment Methodology

The most elementary approach to risk assessment starts with identification of a set of assets $A = \{a_1, a_1, ..., a_n\}$ and threats $T = \{t_1, t_2, ..., t_n\}$. Next, a Cartesian product is formed $A \times T = \{(a_1, t_1), (a_2, t_1), ..., (a_n, t_m)\}$. The value of each asset $v(a_n)$ is determined and, for each threat, the probability of interaction with asset during certain period is assessed $Ea_n(t_m)$. An interaction is problematic only if asset is exposed to vulnerability $Vt_m(a_n) \in [0,1]$. Taking this into account, an appropriate risk estimate is obtained as following.

$$R(a_n, t_m) = v(a_n) * Ea_n(t_m) * Vt_m(a_n) \tag{1}$$

The real problem with this procedure is obtaining exact quantitative values for the above variables in real-time for the following reasons.

- Old statistical data are not available, because the technological landscape and IS change quickly to meet evolving business requirements. Within these changes, new vulnerabilities are created. In addition, different threats are attracted at different time, because business context and assets change over time. Consequently, likelihood of attack and number of vulnerabilities and exposures change over time.
- Furthermore, a substantial proportion of an organization's assets are intangible assets, such as information and goodwill. Identification and valuation of these assets remains a difficult issue [4]. Even worse, the most important asset is personnel. Due to the specifics of this type of assets their valuation is very hard. For example, none of them are recorded and valued in balance sheets.

Therefore, it is hard to derive the exact value of risk. The above facts lead to the current view that the logical alternative to quantitative IS risk assessment is a qualitative approach at the level of aggregates. Here, assets, threats, and vulnerabilities are each categorized intro certain classes. By using tables, such as one below, risks are assessed and estimated, and priorities are set by rank-ordering data on an ordinal scale.

Table 1. The ISO/IEC 27005 risk assessment matrix measures risk on a scale of 0 to 8 and takes two qualitative inputs: (i) likelihood of an incident scenario and (ii) the estimated business impact. For example, if the estimated likelihood of incident scenario is low and the corresponding business impact is high, then the risk is described by the value 4.

		Likelihood of Incident Scenario				
		Very Low	Low	Medium	High	Very High
Business Impact	Very Low	0	1	2	3	4
	Low	1	2	3	4	5
	Medium	2	3	4	5	6
	High	3	4	5	6	7
	Very High	4	5	6	7	8

This is also a legitimate approach according to standards, such as ISO/IEC 27005. However qualitative risk assessment approaches have significant shortcomings and suffer from the following two major disadvantages [3].

- Reversed rankings, i.e., assigning higher qualitative risk ratings to situations that have lower quantitative risks.
- Uninformative ratings, i.e., (i) frequently assigning the most severe qualitative risk label (such as "high") to scenarios with arbitrarily small quantitative risks and (ii) assigning the same ratings to risk that differ by many orders of magnitude.

Therefore, the value of information that qualitative risk assessment approaches provide for improving risk management decision making can be close to zero and misleading in many cases of many small risks and a few large ones, where qualitative ratings often do not distinguish the large risk from the small. This is further justification that quantitative risk treatment has always to be the preferred option, if metrics and measurement methods are available.

5 Metrics and Measurement Problems

The very first activity for successful risk assessment is data collection. These data should include new threats, identified vulnerabilities, exposure times and the available safeguards. This collection and dissemination of data should be in real time to ensure a proactive approach to risk management and also self-adapting security systems. In order to accomplish this, acquisition and distribution process have to be automated.

It needs to be emphasized that, although security in IS has been an important issue for a few decades, there is a lack of appropriate metrics and measurement methods. During measurement activity, numerals are assigned to measures under different rules that lead to different kinds of scales.

Qualitative scale types are nowadays used predominantly for information security measurements. Under Steven's taxonomy [23], they are classified as nominal (categorical) and ordinal. Ordinal scales determine only greater or lesser operation between two measurements. The difference operation between two measurements is not allowed and has no meaning, because successive intervals on the scale are generally unequal in size. Nevertheless, statistical operations such as mean and standard deviation are frequently performed on rank-ordering data, but conclusions drawn from these operations can be misleading.

Instead, quantitative scale types, such as interval and ration scales, should be used to outcome shortcomings described above and consequently, to provide more accurate feedback information. Some important advances have been achieved towards quantitative vulnerability metrics recently. The first such advances are constituted by two databases, the MITRE Corporation Common Vulnerabilities and Exposures [18] and U.S. National Vulnerability Database [19]. These are closely related efforts in which online acquisition and distribution of related data have been enabled by the security content automation protocol SCAP [21]. The main procedure with the first of the databases is as follows.

- The basis is the ID vulnerability, which is an 11-digit number, in which the first three digits are assigned as a candidate value (CAN), the next four denote the year of assignment, and the last four denote the serial number of vulnerability or exposure in that year.
- Once vulnerability is identified in this way, the CAN value is converted to common vulnerability and exposure (CVE).

The data contained in this database are in one of two states. In the first state there are weaknesses with no available patch and in the second state are those variables for which a publicly available patch exists.

This is the basis for the metric called daily vulnerability exposure DVE [14]. DVE is conditional summation formula to calculate how many asset vulnerabilities were public at given date with no corresponding patch and thus possibly leaving a calculated number of assets exposed to threat. DVE values are obtained as follows.

$$DVE(date) = \sum_{vu \ln s} (DATE_{disclosed} < date) \wedge (DATE_{patched} > date) \qquad (2)$$

DVE is useful to show whether an asset is vulnerable and how many vulnerabilities contribute to asset's exposure. In addition, derived DVE trend metric is useful in risk management process to adjust security resources to keep up with rate of disclosed vulnerabilities. Also, additional filtering can be used with DVE such as Common Vulnerability Scoring System (CVSS) [17] to focus on more severe vulnerabilities and exposures. But extra care should be taken when interpreting filtered results, because CVSS filter takes qualitative inputs for quantitative impact evaluation and suffers from same deficiencies as similar qualitative risk assessment approaches.

Another useful metric for our purpose has been proposed by Harriri et al., called the vulnerability index VI [5]. This index is based on categorical assessments of the state of a system: normal, uncertain, or vulnerable. Each network node has an agent that measures the impact factors in real time and sends its reports to a vulnerability analysis engine. The vulnerability analysis engine VAE statistically correlates received data and computes component or system vulnerability and impact metrics. Impact metrics can be used in conjunction with risk evaluation criteria to assess and prioritize risks.

More precise description of VI calculation will be demonstrated for the following fault scenario FS_k. During normal network operation, node's transfer rate is TR_{norm}. Transfer rate may deviate around this value but should not fall below TR_{min}. For each node, CIF is calculated using $TR_{measurement}$.

$$CIF\left(node_j, FS_k\right) = \frac{\left|TR_{norm} - TR_{measurement}\right|}{\left|TR_{norm} - TR_{\min}\right|} \tag{3}$$

Having all CIF values, the component operating state COS can be computed. For each fault scenario FS_k, a operational threshold $d(FS_k)$ is set according to organization's risk acceptance criteria. Next, CIF value is compared to the operational threshold $d(FS_k)$. The resulting COS value equals 1 when the component operates in an abnormal state and 0 when it operates in a normal state.

$$COS\left(node_j, FS_k\right) = \begin{cases} 1, CIF\left(node_j\right) \geq d\left(FS_k\right) \\ 0, CIF\left(node_j\right) < d\left(FS_k\right) \end{cases} \tag{4}$$

Finally, a system impact factor SIF can be computed that identifies how a fault affects the whole network and shows the percentage of components operating in abnormal states, i.e., where CIF exceeds normal operational threshold $d(FS_k)$, in relation to the total number of component. The obtained SIF value can be used can be used in conjunction with risk evaluation criteria to assess and prioritize risks.

$$SIF_{nodes}\left(FS_k\right) = \frac{\sum_{\forall j} COS\left(node_j, FS_k\right)}{totaNumber\, OfNodes} \tag{5}$$

At the end of this sub-section, graph based methods need to be mentioned, which are often used in IS risk assessment and management. One well-established technique in this field is suggested by Schneier [22] and is called attack trees. Attack trees model security of system and subsystems. They support decision making about how to improve security, or evaluate impacts of new attacks. Rote nodes are the principal goals of an attacker and leaf nodes at the bottom represent different attack options.

Intermediate nodes are placed to further refine attacks towards the root node. Nodes can be classified into two categories. When attack can be mounted in many different ways, an "OR" node is used. When attack precondition exists, an "AND" node is used. After the tree is constructed, various metrics can be applied, e.g., cost in time and resources to attack or defend, likelihood of attack and probability of success, or statements about attack such as "Cheapest attack with the highest probability of success".

6 Towards a Computerized Risk Management Architecture

The basis towards computerized risk management is MITRE's initiative called "Making Security Measurable and Manageable" [15]. This initiative has the following structural elements: (i) standardized enumeration of common information security concepts that need to be shared such as vulnerabilities, (ii) languages for encoding and communicating information on common information security concepts, (iii) repositories for sharing these concepts, and (iv) adoption of the above elements by the community through the use of defined program interfaces and their implementations. The initiative is focused on operational level of risk management. Many leading industry security products [20] are already SCAP standard validated, and utilize above described benefits.

Our approach [24] follows the above structure, builds on it (the implementation is still an on-going process), and can be used to improve risk management on strategic and/or tactical level with the development of business flight simulators. In addition, these simulators may be used for automated support of decision-making and could be thus self-adapting security information and event management systems enablers. The IS security environment is complex and it is comprised of information technology and human factor. Because analytical solutions in complex systems are exceptions, we have to rely on computer simulations. Based on this and on the measurement apparatus developed so far, our approach works as follows (see Fig 1).

With regard to achieve efficient risk management, simulations have to be as much realistic as possible. Thus, simulators use two real-time data-feeds.

The US National Vulnerability Database serves as a data-feed for detecting vulnerabilities and exposures of assets. The architecture communicates with the database (or tools provided in the footnote) via SCAP protocol. For example, DVE is calculated by querying the database for each organization's IS asset.

1. Another data-feed is provided through SIF infrastructure where agents monitor the status of nodes in the observed system. Agents and monitoring nodes communicate via SNMP protocol.

In order to successfully correlate organization's asset with newly discovered vulnerabilities and exposures, these have to be registered in NVD. Thus, this data-feed enables reactive risk management. Next data-feed enables proactive risk management, because likelihood and impact of attack is assessed in real-time, regardless if vulnerabilities or exposures are registered in NVD.

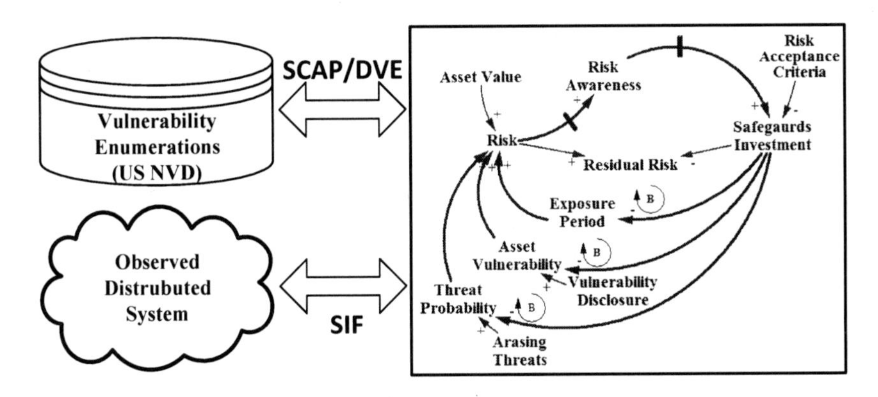

Fig. 1. General Information Technology Risk Management (GIT-RM) model

The numerical representation of acquired data is then used in the simulation model. The aim of simulation is determining dynamics of risk factors to derive information security risk dynamics. The causal dependency of risk factors can be explained in the following way. In the centre of the analysis are assets and threats. Assets are exposed to threats due to their various vulnerabilities. The interaction of threats with vulnerable assets leads to risks. The longer assets are exposed to threats, the higher is the probability of successful exploitation of these vulnerabilities, and therefore the higher the risk. In line with time delayed risk perception, it takes some time to implement appropriate safeguards to reduce exposure period, threat probability and/or asset's vulnerability. After implementation of safeguards according to organization's risk acceptance criteria, some portion of risk may remain effective and is referred to as residual risk.

The mathematical model that formalizes description above [25] is based on first-order differential calculus and is originally provided in Vensim syntax. We present its state-space representation to provide some insight into model characteristics. The internal states, like asset value, are represented by vector $\mathbf{x}(t)$. Vector $\mathbf{u}(t)$ describes inputs, such as arising threats or vulnerability disclosure and $\mathbf{y}(t)$ is output vector to derive residual risk. The system output is defined by functions $\mathbf{h}(t, \mathbf{x}(t), \mathbf{u}(t))$, while the system state can change with respect to current state and its inputs and it is modelled with functions $\mathbf{f}(t, \mathbf{x}(t), \mathbf{u}(t))$. Due to causal dependencies of risk factors and consequently their representation in equations, this system is classified as non-linear. For example, asset vulnerability and safeguard investment states are mutually depended. This system is also time-variant, because it can change in the simulation time, due to PDCA risk management process.

$$\begin{aligned} \frac{d\mathbf{x}}{dt} &= \mathbf{f}\left(t, \mathbf{x}(t), \mathbf{u}(t)\right), \\ \mathbf{y} &= \mathbf{h}\left(t, \mathbf{x}(t), \mathbf{u}(t)\right) \end{aligned} \tag{6}$$

7 Conclusion

User dependence on ICT is rising, primarily because of ever-increasing number of customers that use and rely on electronic services. The value of electronic business is also rising with it and the amount of personal information on Internet is nowadays greater than ever before. Secondarily, the complexity of ICT is increasing. As a consequence, it is more and more difficult to assure security requirements for such systems. Furthermore, some traditional risk assessment techniques are no longer satisfactory for efficient risk management.

These factors together with the fact that exposed assets can be exploited automatically, remotely and with low risk, increasingly attract threat agents in cyberspace. Recent attacks on industrial infrastructure in 2010-2011 revealed that the attacks are becoming more advanced and targeted. Attackers also make great profits because risk management is reactive rather proactive. Thus, risk management paradigm has to change in order to increase odds of a timely risk identification and incident prevention.

In this section, we present a computerized risk management approach that supports active risk management and is architecturally aligned with MITRE's initiative. We argue that quantitative methods should be always preferred over qualitative in order to improve accuracy of risk assessment and efficiency of risk management. Thus, our approach is based on quantitative methods (DVE and SIF) and uses quantitative measurements provided by NVD and monitoring agents in observed system. Self-adapting security information and event management systems will be successful only if accurate information security feedback is provided.

References

1. Bellovin, S.M.: On the Brittleness of Software and the Infeasibility of Security Metrics. IEEE Security & Privacy Magazine 4(4), 96–96 (2006)
2. Centre for Secure Information Technologies: The World Cyber Security Technology Research Summit Report. Belfast (2011)
3. Cox, L.A.T., Babayev, D., Huber, W.: Some limitations of qualitative risk rating systems. Risk Analysis: An Official Publication of the Society for Risk Analysis 25(3), 651–662 (2005)
4. Gerber, M., von Solms, R.: Management of risk in the information age. Computers & Security 24(1), 16–30 (2005)
5. Hariri, S., Dharmagadda, T., Ramkishore, M., Raghavendra, C.S.: Impact analysis of faults and attacks in large-scale networks. IEEE Security & Privacy Magazine 1(5), 49–54 (2003)
6. HIPAA, Basics of Risk Analysis and Risk Management. Washington, USA (2005)
7. ISO/IEC 15408-1:2009, Information technology - Security techniques - Evaluation criteria for IT security - Part 1: Introduction and general model. ISO/IEC (2009)

8. ISO/IEC 21827:2008, Information technology - Security techniques - Systems Security Engineering - Capability Maturity Model (SSE-CMM). ISO/IEC (2008)
9. ISO/IEC 27000:2009, Information technology - Security techniques - Information security management systems - Overview and vocabulary. ISO/IEC (2009)
10. ISO/IEC 27001:2005, Information technology - Security techniques - Information security management systems - Requirements. ISO/IEC (2005)
11. ISO/IEC 27002:2005, Information technology - Security techniques - Code of practice for information security management. ISO/IEC (2005)
12. ISO/IEC 27005:2008, Information technology - Security techniques - Information security risk management. ISO/IEC (2008)
13. ISO/IEC TR 15443-2:2005, Information technology - Security techniques - A framework for IT security assurance - Part 2: Assurance methods (2005)
14. Jones, J.R.: Estimating Software Vulnerabilities. IEEE Security & Privacy Magazine 5(4), 28–32 (2007)
15. Martin, R.A.: Making security measurable and manageable. In: MILCOM 2008 - 2008 IEEE Military Communications Conference, pp. 1–9 (2008)
16. McHugh, J., Fithen, W.L., Arbaugh, W.A.: Windows of vulnerability: a case study analysis. Computer 33(12), 52–59 (2000)
17. Mell, P., Scarfone, K., Romanosky, S.: Common Vulnerability Scoring System. IEEE Security & Privacy Magazine 4(6), 85–89 (2006)
18. MITRE Corp., Common Vulnerabilities and Exposures: The Standard for Information Security Vulnerabilities Names, http://cve.mitre.org/ (accessed: November 19, 2011)
19. NIST, National Vulnerability Database: automating vulnerability management, security measurement, and complience checking, http://nvd.nist.gov/ (accessed: November 19, 2011)
20. NIST, Security Content Automation Protocol Validated Products (2011), http://nvd.nist.gov/scapproducts.cfm (accessed: November 27, 2011)
21. NIST SP 800-126, The Technical Specification for the Security Content Automation Protocol (SCAP): SCAP Version 1.1 (DRAFT), NIST (2010)
22. Schneier, B.: Attack trees. Dr. Dobb's Journal (12), 21–29 (1999)
23. Stevens, S.S.: On the Theory of Scales of Measurement. Science 103(2684), 677–680 (1946)
24. Trček, D.: Security Metrics Foundations for Computer Security. The Computer Journal 53(7), 1106–1112 (2009)
25. Trček, D.: Computationally Supported Quantitative Risk Management for Information Systems. In: Gülpnar, N., Harrison, P., Rüstem, B. (eds.) Performance Models and Risk Management in Communications Systems (Springer Optimization and Its Applications), p. 258. Springer (2010)

Résumé – Summary and Conclusions

Burkhard Stiller

University of Zürich, Communication Systems Group CSG, Switzerland
stiller@ifi.uzh.ch

The field of telecommunications and its relations to economics, regulation, and policies, combined with the technology and operations of large, decentralized networks is an enormous one. In that respect, the goal of the COST Action IS0605 "Econ@Tel" – A Telecommunications Economics COST Network – was to discuss, develop, and outline cross-disciplinary strategic research directions. In the context and final operation of a COST Action, a multitude of different dimensions covered among others:

a) a training network among key people and organizations in order to enhance Europe's competence in the field of telecommunications and media economics,

b) the support of related research and development initiatives, and

c) the provisioning of guidelines and recommendations to European players (such as end-users, enterprises, operators, regulators, policy makers, content providers) for new converged broadband, wireless, and content delivery networks to citizens and enterprises.

The COST Action IS0605 and its partner countries partially coordinated the development of research methodologies as well as tools from engineering, media, and business research, while specifically addressing the identification, study, and discussion of proposed solutions in:

1) evolutionary and regulatory issues that may help or hinder the adoption of economically efficient services,

2) social and policy implications of communications and its technology,

3) economics and governance of future networks, and

4) future network management architectures and related mechanisms.

In that respect, the COST Action IS0605 has mobilized the "critical mass" and diversity of economists, business research experts, engineers, network experts, and scientists working in communications and content economics.

Thus, this book offers to the research community a selection of scientific papers, which over the past four years of the Action's duration have addressed a variety of issues in those aforementioned areas – it does neither intend to be complete nor embracing all cross-disciplinary aspects. These issues presented may seem diverging, and they are, to a certain extend, however almost all of them outline a forward-looking landscape for Telecommunication Economics. As such, the Econ@Tel Action has been in favor of working on mid-term and also long-term aspects of telecommunication economics, of

A.M. Hadjiantonis and B. Stiller (Eds.): Telecommunication Economics, LNCS 7216, pp. 240–242, 2012.

which many are hard to predict in all levels of details. While this fact does not relate only to technology, but it relates to business issues, legislation issues, regulatory issues, as well as behavioral issues of users, this situation was not considered as a drawback, but as the challenge! It was reasonable to expect that many issues would have been resolved during the Action's lifetime, due to very rapid developments in this area. However, several aspects have remained open and provide a good basis for further research.

In that sense, the cooperation of COST Action IS0605 members and countries has seen a new instantiation of expert groups, research teams, and paper writing collaborations, which would not have happened without the COST Action's instrument. Of course, existing partners have been provided the chance to deepen their collaborations. In addition, industrial and regulator contacts have been established (during Econ@Tel's public workshops as well as selected Working Group meetings) and were utilized to discuss real-life problems and emerging fields of interest. However, the major challenge from the beginning – and in some cases lasting until the Action's end – was to find a common basis of terminology, applied models and systems, and applicable evaluation procedures and mechanisms – for all four areas of Econ@Tel members' areas of expertise. Although technologists, engineers, economists, policy makers, and regulators seem to work on the "same subject", their understanding and use of such knowledge is not identical and tools, models, and approaches utilized vary!

Therefore, the joint work undertaken within those four Econ@Tel Working Groups has revealed a common basis and new directions of interests in research to be addressed soon. As such, and in the dimension of individuals and organizations, the networking resources funded by the COST instrument have been invested in a right, successful, and truly European manner.

Explaining the Organization COST

COST – the acronym for European Cooperation in Science and Technology – is the oldest and widest European intergovernmental network for cooperation in research. Established by the Ministerial Conference in November 1971, COST is presently used by the scientific communities of 36 European countries to cooperate in common research projects supported by national funds.

The funds provided by COST – less than 1% of the total value of the projects – support the COST cooperation networks (COST Actions) through which, with € 30 million per year, more than 30,000 European scientists are involved in research having a total value which exceeds € 2 billion per year. This is the financial worth of the European added value which COST achieves.

A "bottom up approach" (the initiative of launching a COST Action comes from the European scientists themselves), "à la carte participation" (only countries interested in the Action participate), "equality of access" (participation is open also to the scientific communities of countries not belonging to the European Union) and "flexible structure" (easy implementation and light management of the research initiatives) are the main characteristics of COST.

As precursor of advanced multidisciplinary research COST has a very important role for the realization of the European Research Area (ERA) anticipating and complementing the activities of the Framework Programmes, constituting a "bridge" towards the scientific communities of emerging countries, increasing the mobility of researchers across Europe and fostering the establishment of "Networks of Excellence" in many key scientific domains such as: Biomedicine and Molecular Biosciences; Food and Agriculture; Forests, their Products and Services; Materials, Physical and Nano-sciences; Chemistry and Molecular Sciences and Technologies; Earth System Science and Environmental Management; Information and Communication Technologies; Transport and Urban Development; Individuals, Societies, Cultures and Health. It covers basic and more applied research and also addresses issues of pre-normative nature or of societal importance.

Further information can be accessed via the Web at the URL:
`http://www.cost.eu`.

Author Index